Proofs Without Words

Exercises in Visual Thinking

Classroom Resource Materials

Classroom Resource Materials is a series of the Mathematical Association of America that is intended to provide supplementary classroom material for students—laboratory exercises, projects, historical information, textbooks with unusual approaches for presenting mathematical ideas, career information, etc.

The Mathematical Association of America
1529 Eighteenth Street, NW
Washington, DC 20036
1-800-331-1MAA   FAX 202-265-2384

Classroom Resource Materials

Number 1

Proofs Without Words

Exercises in Visual Thinking

Roger B. Nelsen

Lewis and Clark College

Published and Distributed by

THE MATHEMATICAL ASSOCIATION OF AMERICA

©1993 by
The Mathematical Association of America (Incorporated)
Library of Congress Catalog Card Number 93-86388

ISBN 0-88385-700-6

Printed in the United States of America

Current Printing (last digit):
10 9 8 7 6 5 4 3 2 1

Introduction

see (sē) *v.*, **saw, seen, seeing.** —*v.t.*

. . .

5. to perceive (things) mentally; discern;
understand: *to see the point of an argument.*

. . .

      —THE RANDOM HOUSE DICTIONARY
      OF THE ENGLISH LANGUAGE (2nd ED.)
      UNABRIDGED.

"Proofs without words" (PWWs) have become regular features in the journals published by the Mathematical Association of America — notably *Mathematics Magazine* and *The College Mathematics Journal.* PWWs began to appear in *Mathematics Magazine* about 1975, and, in an editors' note in the January 1976 issue of the *Magazine,* J. Arthur Seebach and Lynn Arthur Steen encouraged further contributions of PWWs to the *Magazine.* Although originally solicited for "use as end-of-article fillers," the editors went on to ask "What could be better for this purpose than a pleasing illustration that made an important mathematical point?"

A few years earlier Martin Gardner, in his popular "Mathematical Games" column in the October 1973 issue of the *Scientific American,* discussed PWWs as "look-see" diagrams. Gardner points out that "in many cases a dull proof can be supplemented by a geometric analogue so simple and beautiful that the truth of a theorem is almost seen at a glance." This dramatically illustrates the dictionary quote above: in English "to see" is often "to understand."

In the same vein, the editorial policy of *The College Mathematics Journal* throughout most of the 1980s stated that, in addition to expository articles, "The Journal also invites other types of contributions, most notably: *proofs without words,* mathematical poetry, quotes, ..." (their italics). But PWWs are not recent innovations — they have a long history. Indeed, in this volume you will find modern renditions of proofs without words from ancient China, classical Greece, and India of the twelfth century.

Of course, "proofs without words" are not really proofs. As Theodore Eisenberg and Tommy Dreyfus note in their paper "On the Reluctance to Visualize in Mathematics" [in *Visualization in Teaching and Learning Mathematics,* MAA Notes Number 19], some consider such visual arguments to be of little value, and "that there is one and only one way to communicate mathematics, and 'proofs without words' are not acceptable." But to counter this viewpoint, Eisenberg and Dreyfus go on to give us some quotes on the subject:

> [Paul] Halmos, speaking of Solomon Lefschetz (editor of the *Annals*), stated: "He saw mathematics not as logic but as pictures." Speaking of what it takes to be a mathematician, he stated: "To be a scholar of mathematics you must be born with ... the ability to visualize" and most teachers try to develop this ability in their students. [George] Pólya's "Draw a figure ..." is classic pedagogic advice, and Einstein and Poincaré's views that we should use our visual intuitions are well known.

So, if "proofs without words" are not proofs, what are they? As you will see from this collection, this question does not have a simple, concise answer. But generally, PWWs are pictures or diagrams that help the observer see *why* a particular statement may be true, and also to see *how* one might begin to go about proving it true. In some an equation or two may appear in order to guide the observer in this process. But the emphasis is clearly on providing visual clues to the observer to stimulate mathematical thought.

I should note that this collection is not intended to be complete. It does not include all PWWs which have appeared in print, but is rather a sample representative of the genre. In addition, as readers of the Association's journals are well aware, new PWWs appear in print rather frequently, and I anticipate that this will continue. Perhaps some day a second volume of PWWs will appear!

I hope that the readers of this collection will find enjoyment in discovering or rediscovering some elegant visual demonstrations of certain mathematical ideas; that teachers will want to share many of them with their students; and that all will find stimulation and encouragement to try to create new "proofs without words."

Acknowledgment. I would like to express my appreciation and gratitude to the many people who have played a part in the publication of this collection: to Gerald Alexanderson and Martha Siegel, who, as editors of *Mathematics Magazine*, gave me encouragement over the years as I learned to read and write PWWs; to Doris Schattschneider, Eugene Klotz, and Richard Guy for sharing with me their collections of PWWs; and finally, to all those individuals who have contributed "proofs without words" to the mathematical literature (see the *Index of Names* on pp. 151-152), without whom this collection simply would not exist.

Note. All the drawings in this collection were redone to create a uniform appearance. In a few instances titles were changed, and shading or symbols were added (or deleted) for clarity. Any errors resulting from that process are entirely my responsibility.

Roger B. Nelsen
Lewis and Clark College
Portland, Oregon

Contents

Geometry & Algebra

The Pythagorean Theorem I

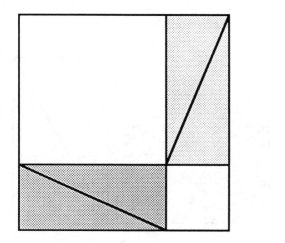 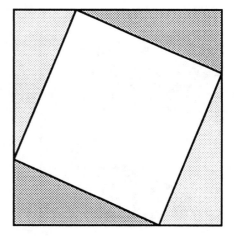

—adapted from the *Chou pei suan ching*
(author unknown, circa B.C. 200?)

The Pythagorean Theorem II

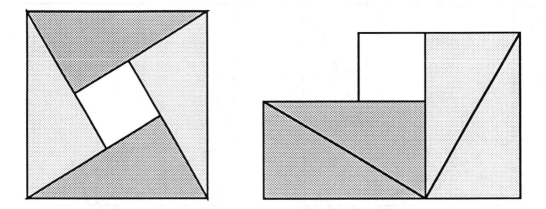

Behold!

—Bhāskara (12th century)

The Pythagorean Theorem III

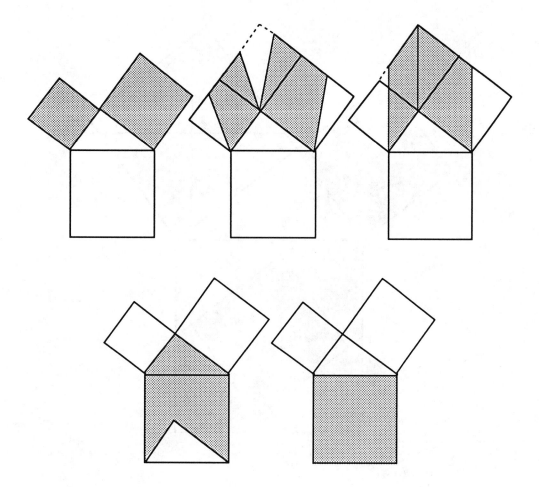

—based on Euclid's proof

The Pythagorean Theorem IV

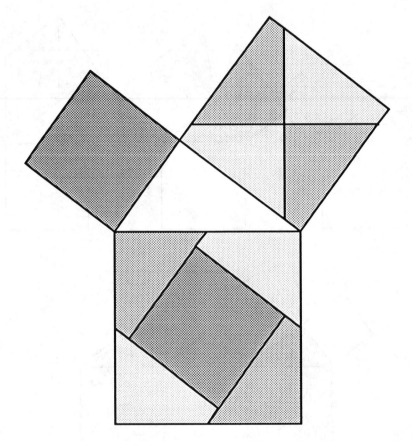

—H. E. Dudeney (1917)

The Pythagorean Theorem V

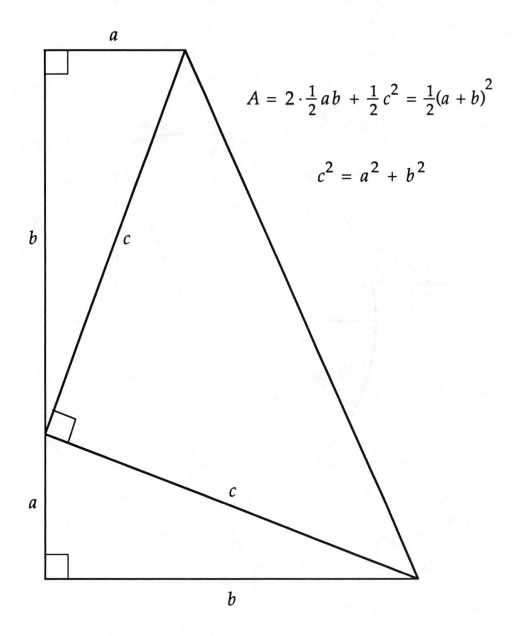

$$A = 2 \cdot \tfrac{1}{2} a b + \tfrac{1}{2} c^2 = \tfrac{1}{2}(a + b)^2$$

$$c^2 = a^2 + b^2$$

—James A. Garfield (1876)
20th President of the United States

The Pythagorean Theorem VI

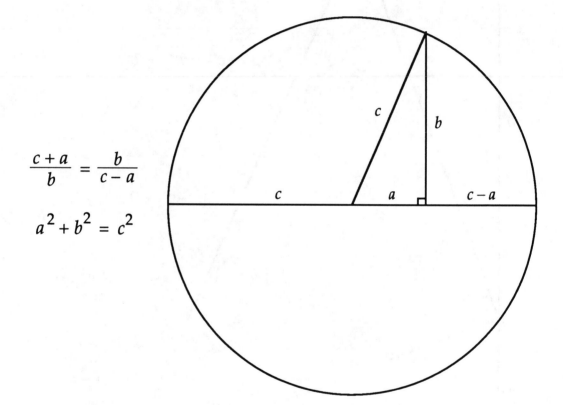

$$\frac{c+a}{b} = \frac{b}{c-a}$$

$$a^2 + b^2 = c^2$$

—Michael Hardy

A Pythagorean Theorem: $a \cdot a' = b \cdot b' + c \cdot c'$

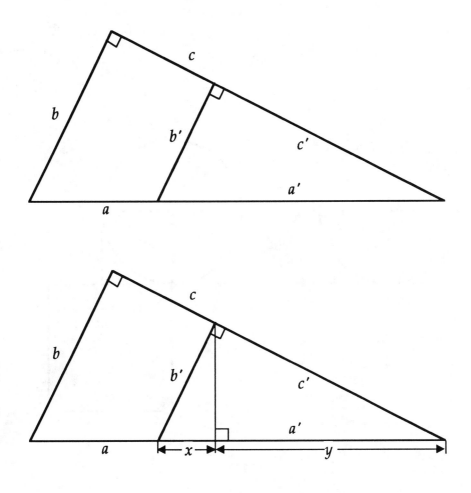

$$\frac{x}{b} = \frac{b'}{a} \implies a \cdot x = b \cdot b';$$

$$\frac{y}{c} = \frac{c'}{a} \implies a \cdot y = c \cdot c';$$

$$\therefore \; a \cdot a' = a \cdot (x + y) = b \cdot b' + c \cdot c'.$$

—Enzo R. Gentile

The Rolling Circle Squares Itself

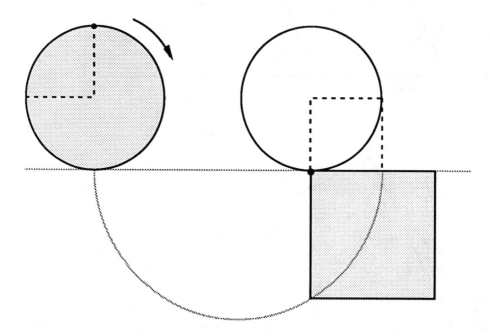

—Thomas Elsner

On Trisecting an Angle

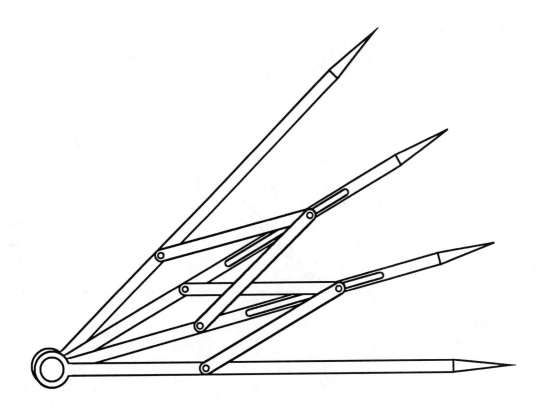

—Rufus Isaacs

Trisection of an Angle in an Infinite Number of Steps

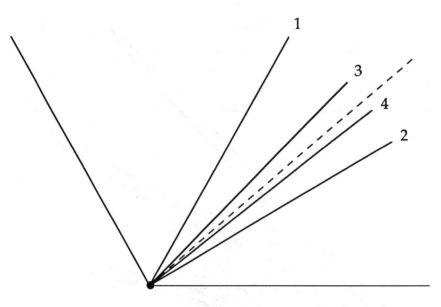

$$\frac{1}{3} = \frac{1}{2} - \frac{1}{4} + \frac{1}{8} - \frac{1}{16} + \cdots$$

—Eric Kincanon

Trisection of a Line Segment

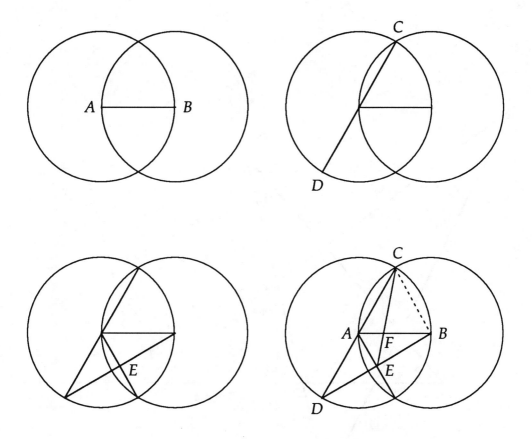

$$\overline{AF} = \frac{1}{3} \cdot \overline{AB}$$

—Scott Coble

The Vertex Angles of a Star Sum to 180°

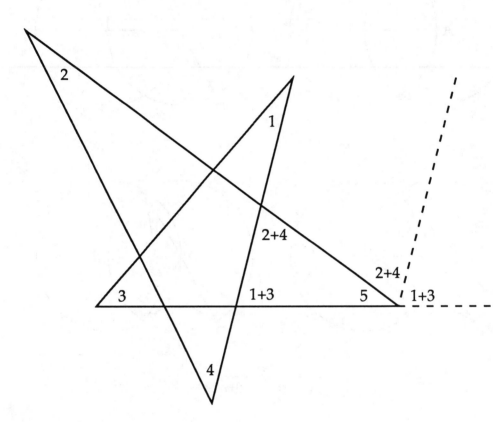

—Fouad Nakhli

Viviani's Theorem

The perpendiculars to the sides from a point on the boundary
or within an equilateral triangle add up to the height
of the triangle.

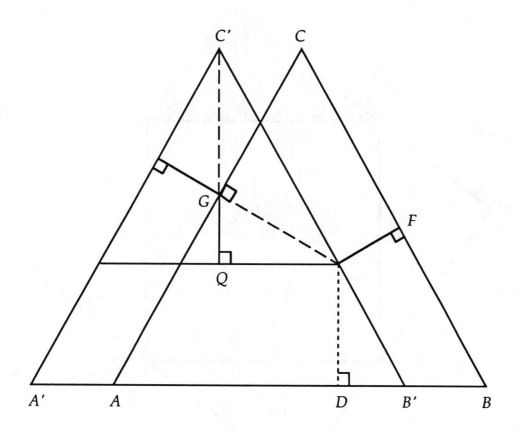

—Samuel Wolf

A Theorem About Right Triangles

The internal bisector of the right angle of a right triangle
bisects the square on the hypotenuse.

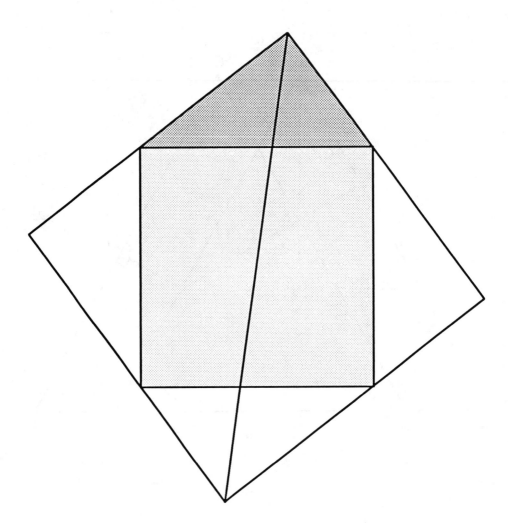

—Roland H. Eddy

Area and the Projection Theorem of a Right Triangle

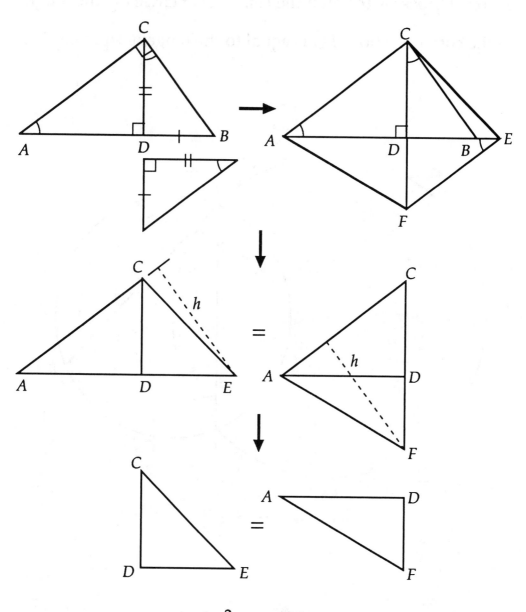

$$CD^2 = AD \cdot DB$$

—Sidney H. Kung

Chords and Tangents of Equal Length

If circle C_1 passes through the center O of circle C_2, the length of the common chord $\overline{PQ}$ is equal to the tangent segment $\overline{PR}$.

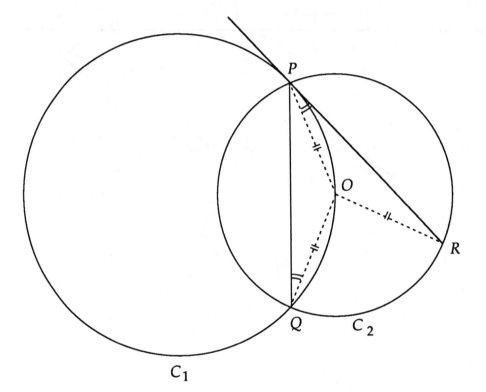

—Roland H. Eddy

Completing the Square

$$x^2 + ax = (x + a/2)^2 - (a/2)^2$$

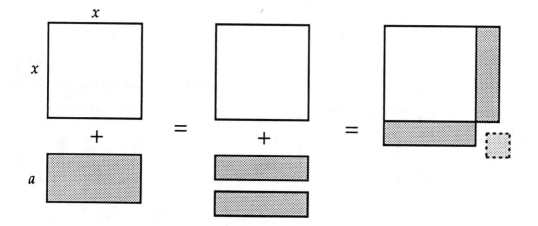

—Charles D. Gallant

Algebraic Areas I

$$(a + b)^2 + (a - b)^2 = 2(a^2 + b^2)$$

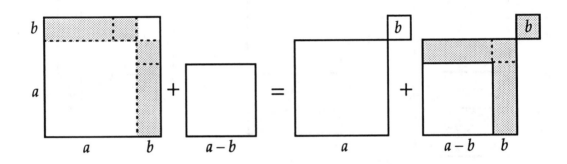

—Shirley Wakin

Algebraic Areas II

$$(a + b + c)^2 + (a + b - c)^2 + (a - b + c)^2 + (a - b - c)^2$$
$$= (2a)^2 + (2b)^2 + (2c)^2$$

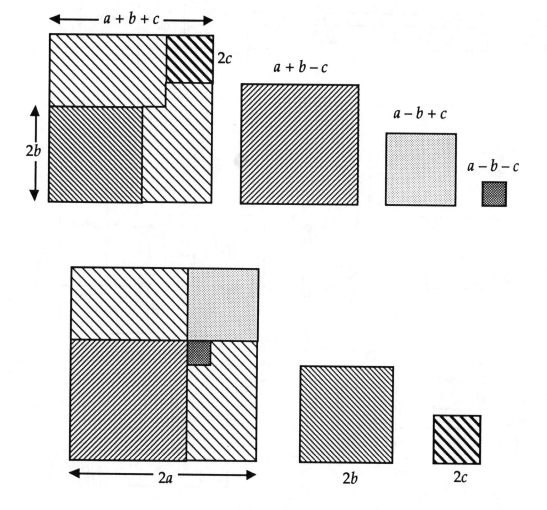

—Sam Pooley and K. Ann Drude

Diophantus of Alexandria's "Sum of Squares" Identity

$$(a^2 + b^2)(c^2 + d^2) = (ad + bc)^2 + (bd - ac)^2$$

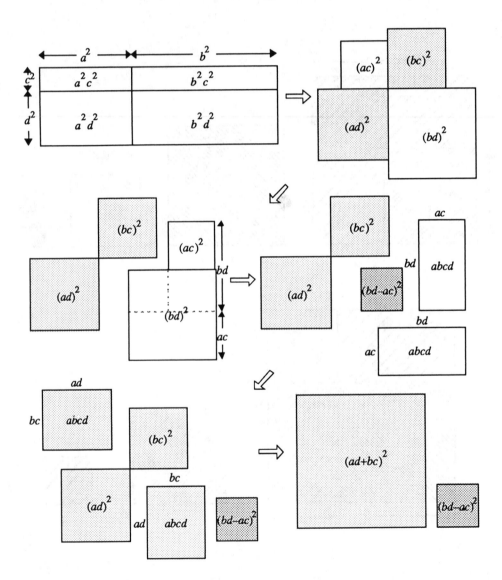

—RBN

The kth n-gonal Number is

$$1 + (k-1)(n-1) + \frac{1}{2}(k-2)(k-1)(n-2)$$

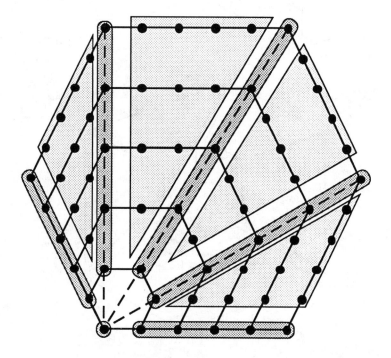

—Dave Logothetti

The Volume of a Frustum of a Square Pyramid
[Problem 14, *The Moscow Papyrus*, circa 1850 B.C.]

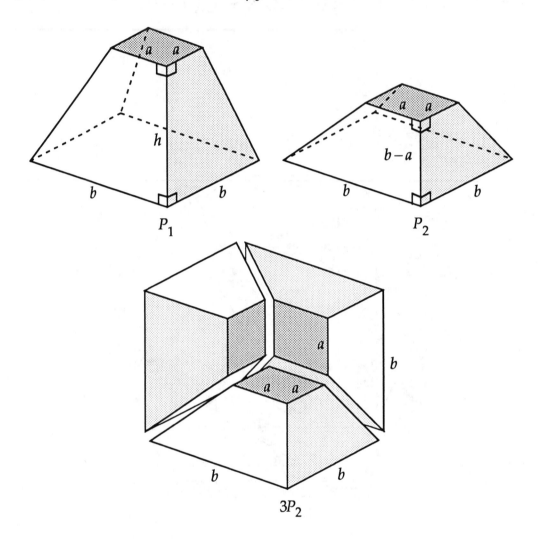

$$V(P_1) = \frac{h}{b-a} \, V(P_2) = \frac{h}{b-a} \cdot \frac{1}{3}\,(b^3 - a^3) = \frac{h}{3}\,(a^2 + ab + b^2)$$

REFERENCES
1. C. B. Boyer, *A History of Mathematics,* John Wiley & Sons, New York, 1968, pp. 20-22.
2. R. J. Gillings, *Mathematics in the Time of the Pharaohs,* The MIT Press, Cambridge, 1972, pp. 187-193.

—RBN

The Volume of a Hemisphere via Cavalieri's Principle*

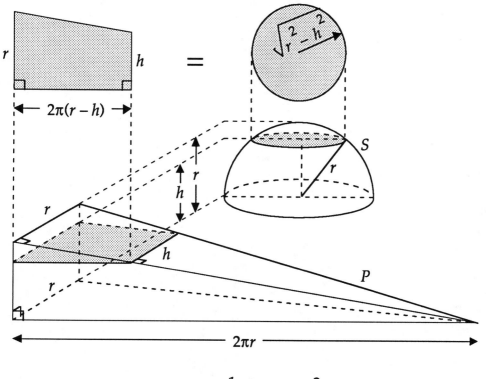

$$V_S = V_P = \frac{1}{3}r^2 \cdot 2\pi r = \frac{2}{3}\pi r^3$$

*Tzu Geng, son of the most celebrated mathematician Tzu Chung Chih in ancient China, was believed to be the first to develop the principle in the 5th century A. D.

—Sidney H. Kung

Trigonometry, Calculus & Analytic Geometry

Sine of the Sum

$$\sin (x + y) = \sin x \cos y + \cos x \sin y \quad \text{for } x + y < \pi$$

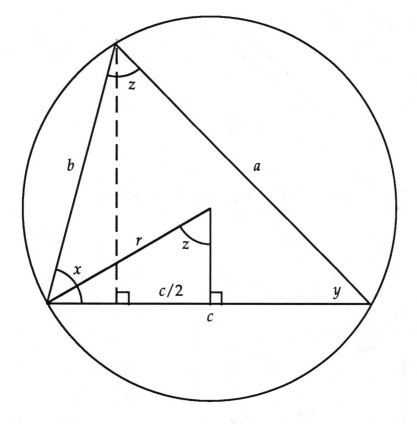

$$c = a \cos y + b \cos x$$

$$r = 1/2 \Rightarrow \sin z = (c/2)/(1/2) = c, \ \sin x = a, \ \sin y = b;$$

$$\sin (x + y) = \sin (\pi - (x + y)) = \sin z = \sin x \cos y + \sin y \cos x$$

—Sidney H. Kung

Area and Difference Formulas

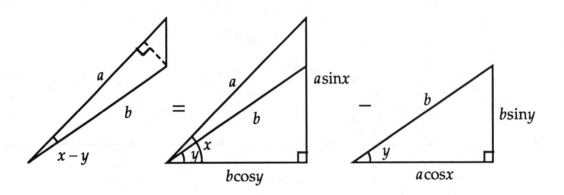

$$\sin(x - y) = \sin x \cos y - \cos x \sin y$$

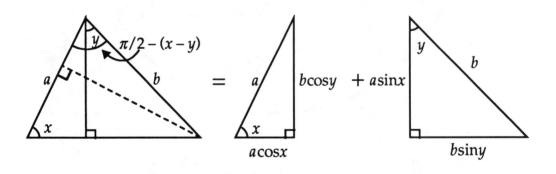

$$\cos(x - y) = \cos x \cos y + \sin x \sin y$$

—Sidney H. Kung

The Law of Cosines I

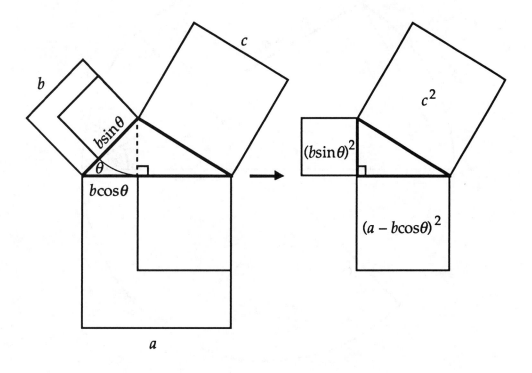

$$c^2 = (b \sin \theta)^2 + (a - b \cos \theta)^2$$
$$= a^2 + b^2 - 2ab \cos \theta$$

—Timothy A. Sipka

The Law of Cosines II

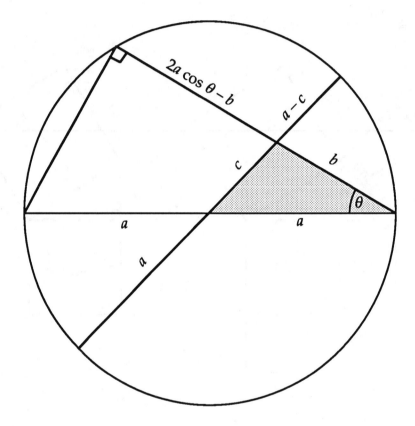

$$(2a \cos \theta - b)b = (a - c)(a + c)$$
$$c^2 = a^2 + b^2 - 2ab \cos \theta$$

—Sidney H. Kung

The Law Of Cosines III (via Ptolemy's Theorem)

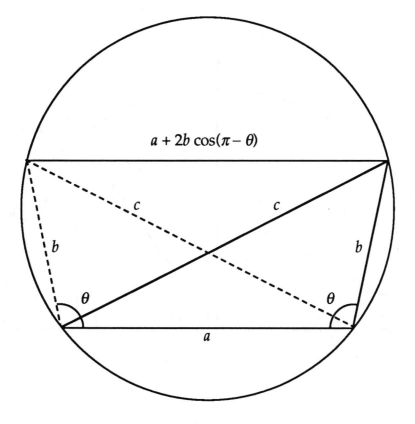

$$c \cdot c = b \cdot b + (a + 2b \cos(\pi - \theta)) \cdot a$$

$$c^2 = a^2 + b^2 - 2ab \cdot \cos\theta$$

—Sidney H. Kung

The Double-Angle Formulas

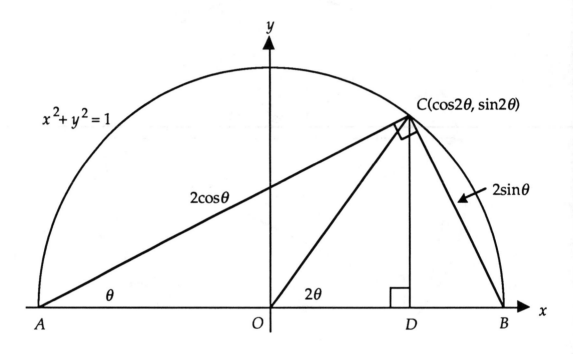

$$\triangle ACD \sim \triangle ABC$$

$$\overline{CD} / \overline{AC} = \overline{BC} / \overline{AB} \qquad \overline{AD} / \overline{AC} = \overline{AC} / \overline{AB}$$

$$\sin 2\theta / 2\cos\theta = 2\sin\theta / 2 \qquad (1 + \cos 2\theta)/2\cos\theta = 2\cos\theta / 2$$

$$\sin 2\theta = 2\sin\theta\cos\theta \qquad \cos 2\theta = 2\cos^2\theta - 1$$

—RBN

The Half-Angle Tangent Formulas

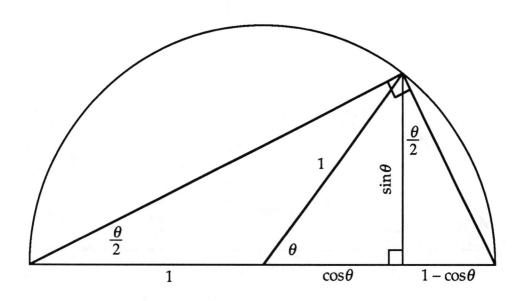

$$\tan \frac{\theta}{2} = \frac{\sin\theta}{1 + \cos\theta} = \frac{1 - \cos\theta}{\sin\theta}$$

—R. J. Walker

Mollweide's Equation

$$(a - b) \cos \frac{\gamma}{2} = c \sin \left(\frac{\alpha - \beta}{2} \right)$$

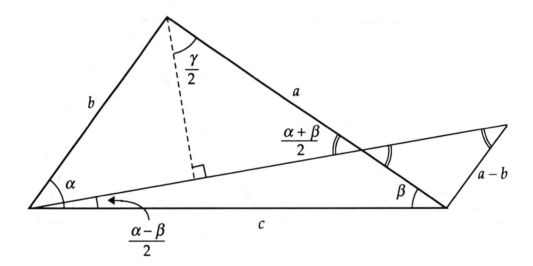

—H. Arthur DeKleine

$$(\tan\theta + 1)^2 + (\cot\theta + 1)^2 = (\sec\theta + \csc\theta)^2$$

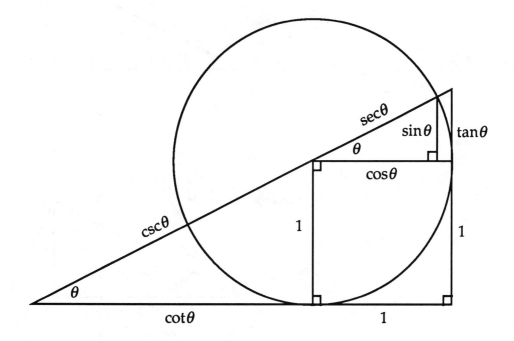

$$\tan^2\theta + 1 = \sec^2\theta$$
$$\cot^2\theta + 1 = \csc^2\theta$$
$$(\tan\theta + 1)^2 + (\cot\theta + 1)^2 = (\sec\theta + \csc\theta)^2$$
$$\left(\text{also } \tan\theta = \frac{\tan\theta + 1}{\cot\theta + 1} \right)$$

—William Romaine

The Substitution to Make a Rational Function of the Sine and Cosine

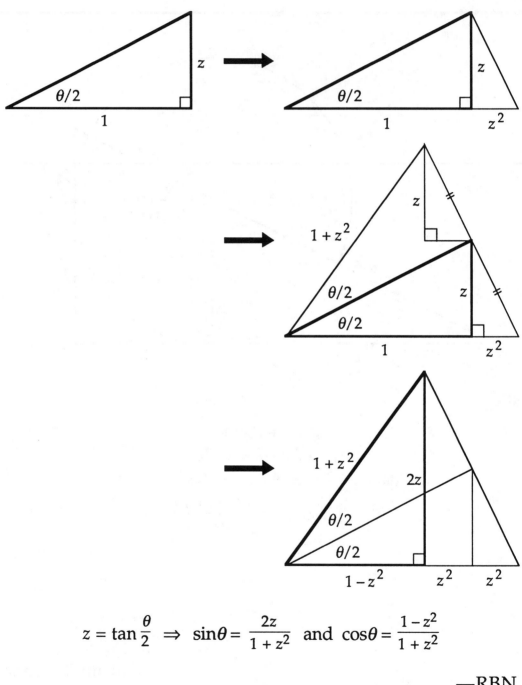

$$z = \tan\frac{\theta}{2} \;\Rightarrow\; \sin\theta = \frac{2z}{1+z^2} \;\text{ and }\; \cos\theta = \frac{1-z^2}{1+z^2}$$

—RBN

Sums of Arctangents

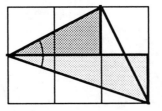

$$\arctan \frac{1}{2} + \arctan \frac{1}{3} = \frac{\pi}{4}$$

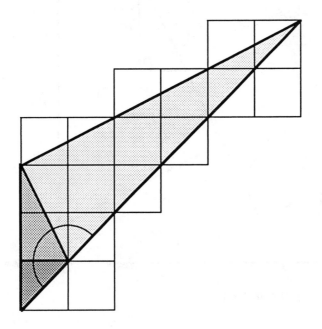

$$\arctan 1 + \arctan 2 + \arctan 3 = \pi$$

—Edward M. Harris

The Distance Between a Point and a Line

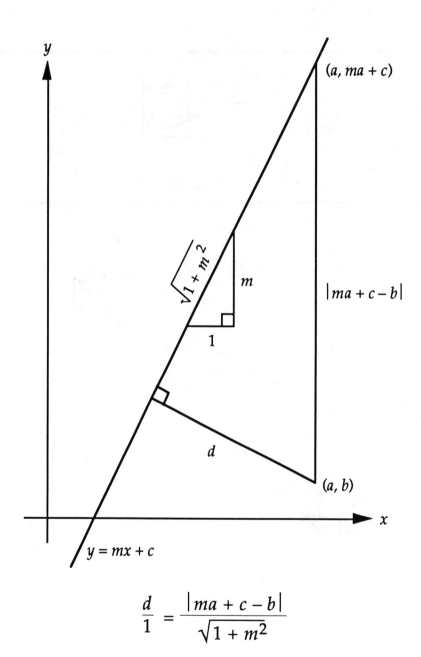

$$\frac{d}{1} = \frac{|ma + c - b|}{\sqrt{1 + m^2}}$$

—R. L. Eisenman

The Midpoint Rule is Better than the Trapezoidal Rule for Concave Functions

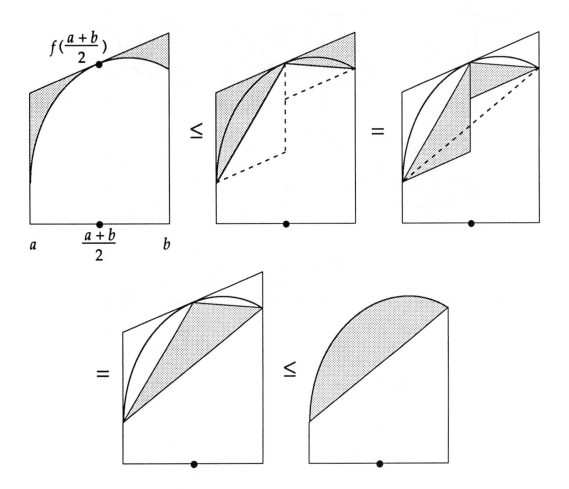

—Frank Burk

Integration by Parts

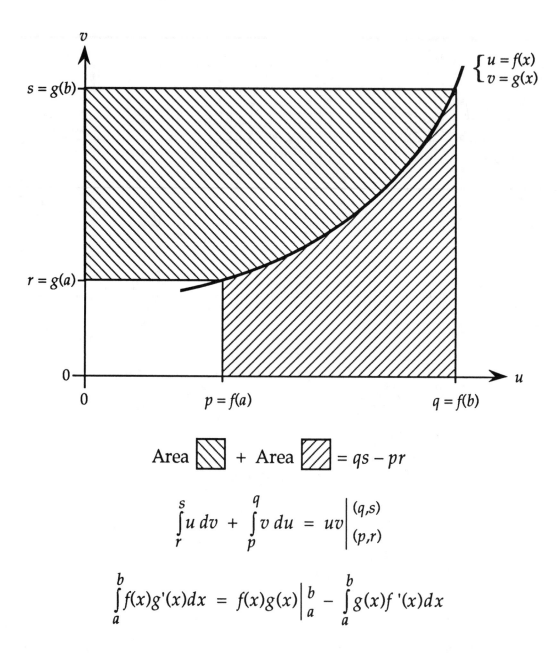

$$\text{Area } \diagdown + \text{ Area } \diagup = qs - pr$$

$$\int_r^s u\,dv + \int_p^q v\,du = uv\bigg|_{(p,r)}^{(q,s)}$$

$$\int_a^b f(x)g'(x)dx = f(x)g(x)\bigg|_a^b - \int_a^b g(x)f'(x)dx$$

—Richard Courant

The Graphs of f and f^{-1} are Reflections about the Line $y = x$

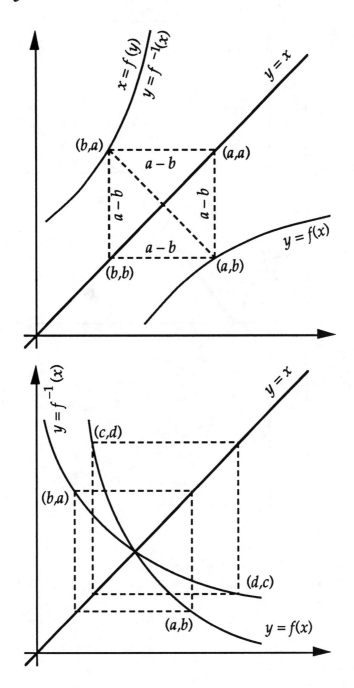

—Ayoub B. Ayoub

The Reflection Property of the Parabola

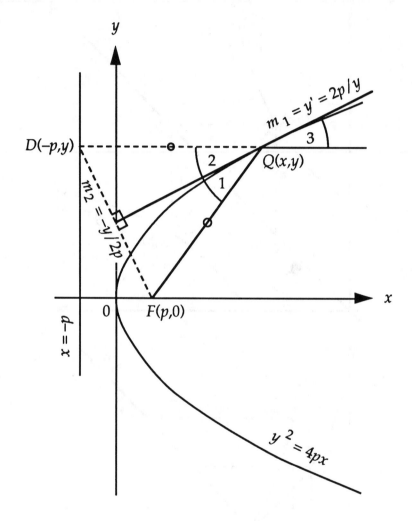

$$QF = QD \quad \& \quad m_1 \cdot m_2 = -1 \quad \Rightarrow \quad \angle 1 = \angle 2 = \angle 3$$

—Ayoub B. Ayoub

Area under an Arch of the Cycloid

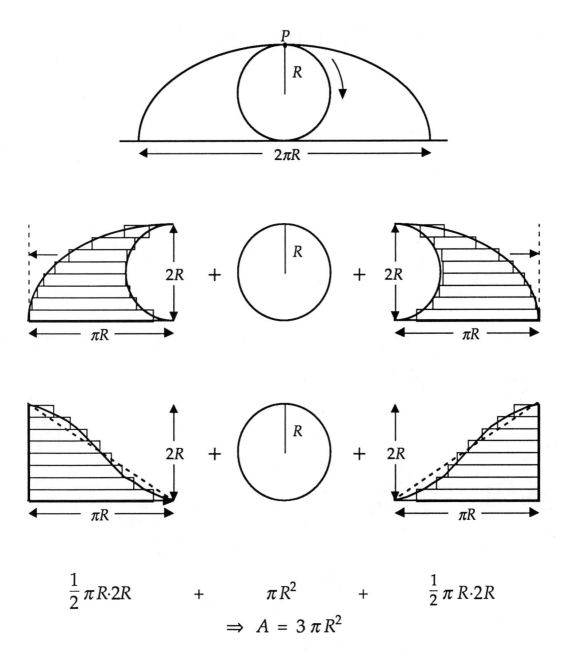

$$\frac{1}{2}\pi R \cdot 2R \quad + \quad \pi R^2 \quad + \quad \frac{1}{2}\pi R \cdot 2R$$

$$\Rightarrow A = 3\pi R^2$$

—Richard M. Beekman

Inequalities

The Arithmetic Mean—Geometric Mean Inequality I

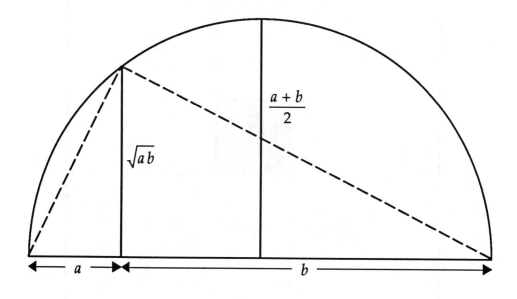

$$\sqrt{ab} \leq \frac{a+b}{2}$$

—Charles D. Gallant

The Arithmetic Mean—Geometric Mean Inequality II

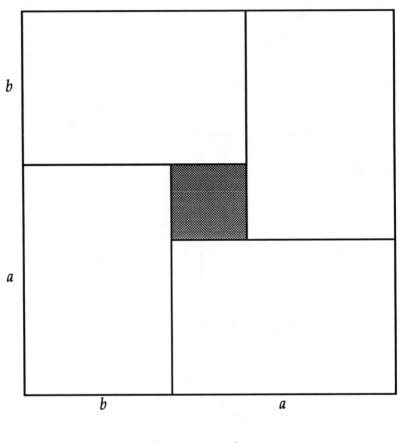

$$(a + b)^2 - (a - b)^2 = 4ab$$

$$\frac{a + b}{2} \geq \sqrt{ab}$$

—Doris Schattschneider

The Arithmetic Mean—Geometric Mean Inequality III

$$\frac{a+b}{2} \geq \sqrt{ab}, \text{ with equality if and only if } a = b$$

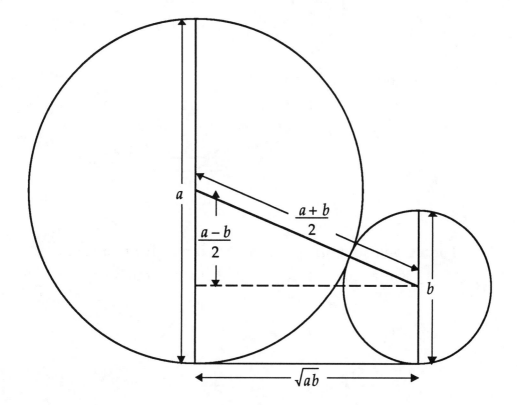

—Roland H. Eddy

Two Extremum Problems

For a given product, the sum of two positive numbers
is minimal when the numbers are equal.

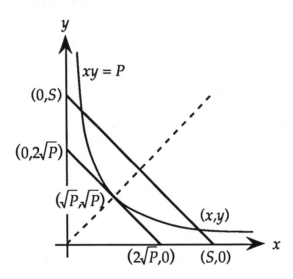

For a given sum, the product of two positive numbers
is maximal when the numbers are equal.

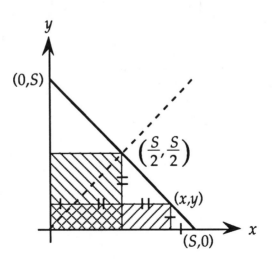

—Paolo Montuchi and Warren Page

The Harmonic Mean—Geometric Mean—Arithmetic Mean—Root Mean Square Inequality I

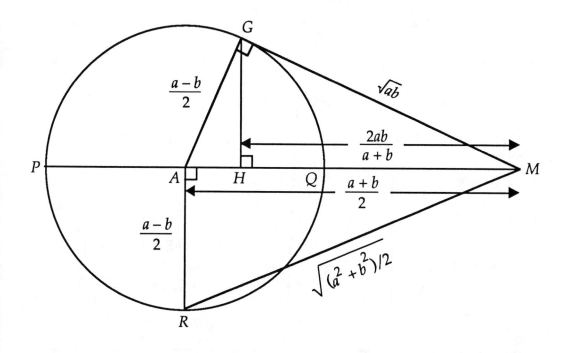

$$PM = a, \quad QM = b, \quad a > b > 0$$

$$HM < GM < AM < RM$$

$$\frac{2ab}{a+b} < \sqrt{ab} < \frac{a+b}{2} < \sqrt{(a^2 + b^2)/2}$$

—RBN

The Harmonic Mean—Geometric Mean—Arithmetic Mean—Root Mean Square Inequality II

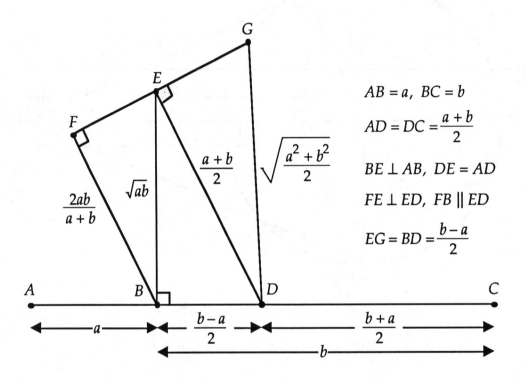

$AB = a$, $BC = b$

$AD = DC = \dfrac{a+b}{2}$

$BE \perp AB$, $DE = AD$

$FE \perp ED$, $FB \parallel ED$

$EG = BD = \dfrac{b-a}{2}$

—Sidney H. Kung

The Harmonic Mean—Geometric Mean—Arithmetic Mean—Root Mean Square Inequality III

$$a, b > 0 \Rightarrow \sqrt{(a^2 + b^2)/2} \geq \frac{a + b}{2} \geq \sqrt{ab} \geq \frac{2ab}{a + b}$$

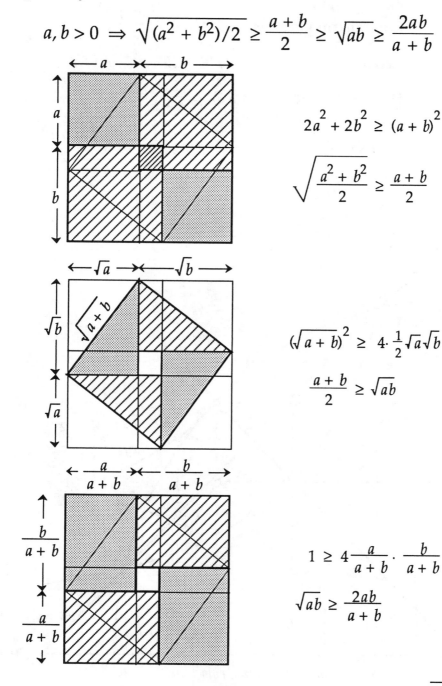

$$2a^2 + 2b^2 \geq (a + b)^2$$

$$\sqrt{\frac{a^2 + b^2}{2}} \geq \frac{a + b}{2}$$

$$(\sqrt{a + b})^2 \geq 4 \cdot \frac{1}{2}\sqrt{a}\sqrt{b}$$

$$\frac{a + b}{2} \geq \sqrt{ab}$$

$$1 \geq 4\frac{a}{a + b} \cdot \frac{b}{a + b}$$

$$\sqrt{ab} \geq \frac{2ab}{a + b}$$

—RBN

Five Means — and Their Means

Arithmetic: $am = AM(a,b) = \dfrac{a+b}{2}$

Contraharmonic: $cm = CM(a,b) = \dfrac{a^2 + b^2}{a+b}$

Geometric: $gm = GM(a,b) = \sqrt{ab}$

Harmonic: $hm = HM(a,b) = \dfrac{2ab}{a+b}$

Root Mean Square: $rms = RMS(a,b) = \sqrt{\dfrac{a^2 + b^2}{2}}$

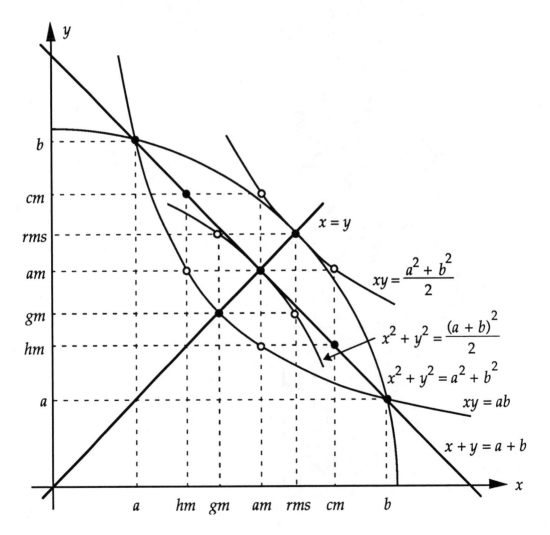

I. $0 < a < b \Rightarrow$

$$a \;<\; \frac{2ab}{a+b} \;<\; \sqrt{ab} \;<\; \frac{a+b}{2} \;<\; \sqrt{\frac{a^2+b^2}{2}} \;<\; \frac{a^2+b^2}{a+b} \;<\; b$$

II. $hm + cm = a + b \Rightarrow AM(hm, cm) = am.$

III. $hm \cdot am = a \cdot b \Rightarrow GM(hm, am) = gm.$

IV. $am \cdot cm = \dfrac{a^2 + b^2}{2} \Rightarrow GM(am, cm) = rms.$

V. $gm^2 + rms^2 = \dfrac{(a+b)^2}{2} \Rightarrow RMS(gm, rms) = am.$

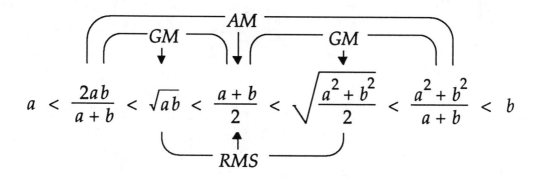

—RBN

$$e^{\pi} > \pi^e$$

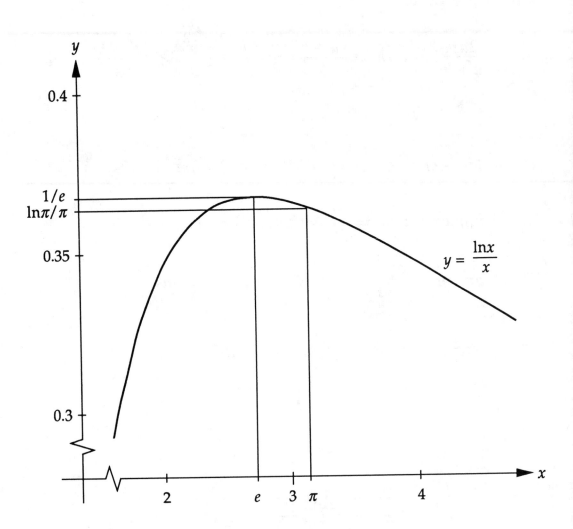

—Fouad Nakhli

$$A^B > B^A \text{ for } e \leq A < B$$

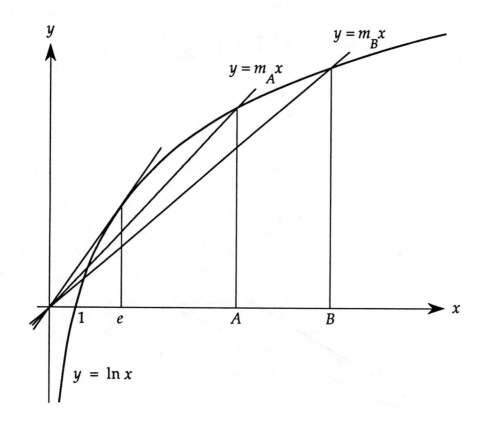

$$e \leq A < B \implies m_A > m_B$$

$$\implies \frac{\ln A}{A} > \frac{\ln B}{B}$$

$$\implies A^B > B^A$$

—Charles D. Gallant

The Mediant Property

$$\frac{a}{b} < \frac{c}{d} \;\Rightarrow\; \frac{a}{b} < \frac{a+c}{b+d} < \frac{c}{d}$$

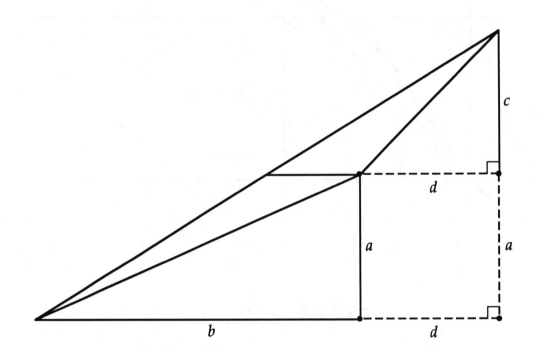

—Richard A. Gibbs

Regle des Nombres Moyens (Two Proofs)
[Nicolas Chuquet, *Le Triparty en la Science des Nombres*, 1484]

$$a, b, c, d > 0; \quad \frac{a}{b} < \frac{c}{d} \implies \frac{a}{b} < \frac{a+c}{b+d} < \frac{c}{d}$$

I.

$$m_1 < m_3 \implies m_1 < m_2 < m_3$$

—Li Changming

II.

$$\frac{a}{b} < \frac{a}{b+d} + \frac{c}{b+d} < \frac{c}{d}$$

—RBN

The Sum of a Positive Number and its Reciprocal is at least Two (Four Proofs)

I.

II.

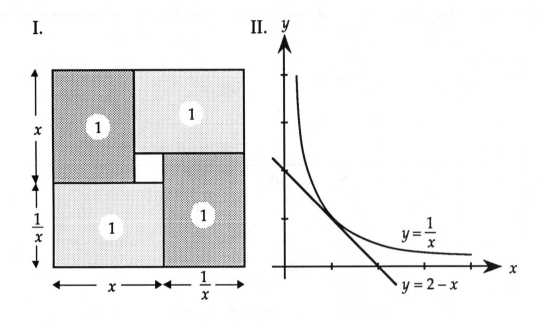

III.

IV.

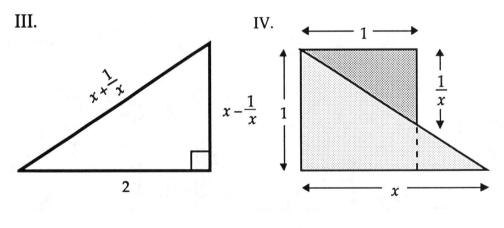

$$x \geq 1 \;\Rightarrow\; x + \frac{1}{x} \geq 2$$

—RBN

Aristarchus' Inequalities

$$0 < \beta < \alpha < \frac{\pi}{2} \Rightarrow \frac{\sin\alpha}{\sin\beta} < \frac{\alpha}{\beta} < \frac{\tan\alpha}{\tan\beta}$$

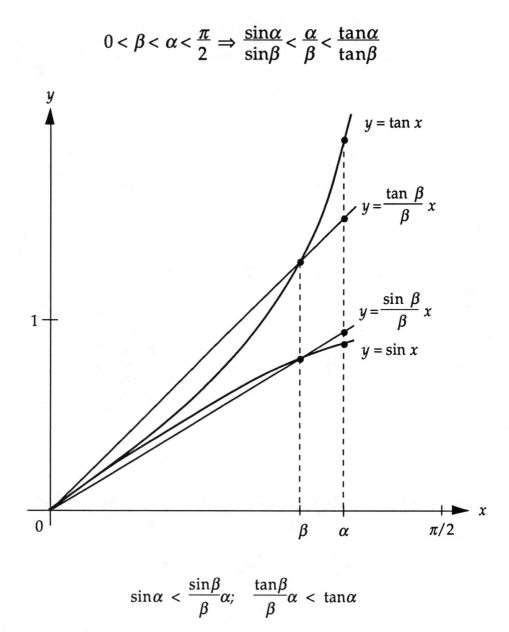

$$\sin\alpha < \frac{\sin\beta}{\beta}\alpha; \quad \frac{\tan\beta}{\beta}\alpha < \tan\alpha$$

$$\therefore \ \frac{\sin\alpha}{\sin\beta} < \frac{\alpha}{\beta} < \frac{\tan\alpha}{\tan\beta}$$

—RBN

The Cauchy-Schwarz Inequality

$$|\langle a,b \rangle \cdot \langle x,y \rangle| \le \|\langle a,b \rangle\| \; \|\langle x,y \rangle\|$$

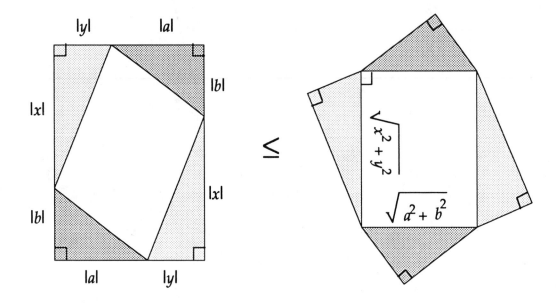

$$(|a| + |y|)(|b| + |x|) \le 2(\tfrac{1}{2}|a||b| + \tfrac{1}{2}|x||y|) + \sqrt{a^2 + b^2}\sqrt{x^2 + y^2}$$

$$\therefore |ax + by| \le |a||x| + |b||y| \le \sqrt{a^2 + b^2}\sqrt{x^2 + y^2}$$

—RBN

Bernoulli's Inequality (two proofs)

$$x > 0, x \neq 1, r > 1 \implies x^r - 1 > r(x - 1)$$

I. (first semester calculus)

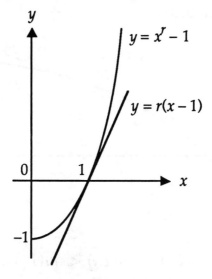

II. (second semester calculus)

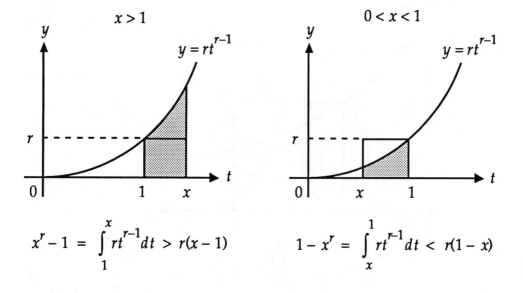

$$x^r - 1 = \int_1^x rt^{r-1}dt > r(x - 1) \qquad 1 - x^r = \int_x^1 rt^{r-1}dt < r(1 - x)$$

—RBN

Napier's Inequality (two proofs)

$$b > a > 0 \ \Rightarrow \ \frac{1}{b} < \frac{\ln b - \ln a}{b - a} < \frac{1}{a}$$

I. (first semester calculus)

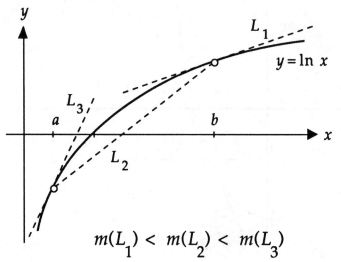

$$m(L_1) \ < \ m(L_2) \ < \ m(L_3)$$

II. (second semester calculus)

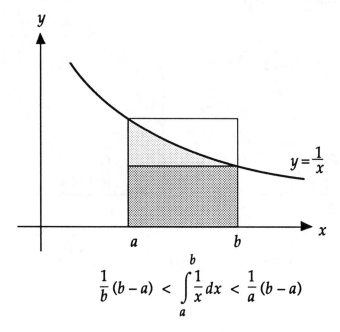

$$\frac{1}{b}(b - a) \ < \ \int_a^b \frac{1}{x}\,dx \ < \ \frac{1}{a}(b - a)$$

—RBN

Integer Sums

Sums of Integers I

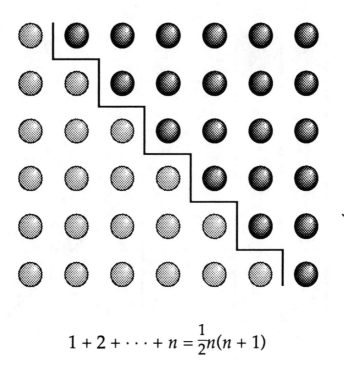

$$1 + 2 + \cdots + n = \frac{1}{2}n(n + 1)$$

—"The ancient Greeks"
(as cited by Martin Gardner)

Sums of Integers II

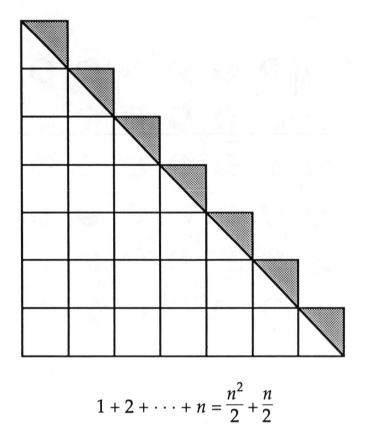

$$1 + 2 + \cdots + n = \frac{n^2}{2} + \frac{n}{2}$$

—Ian Richards

Sums of Odd Integers I

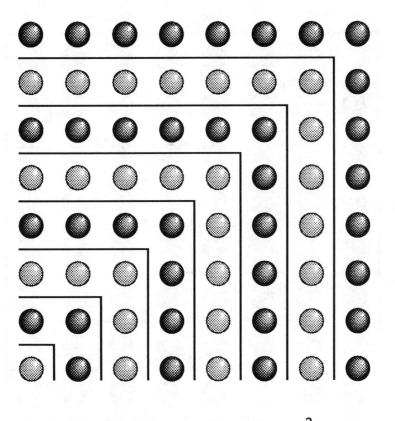

$$1 + 3 + 5 + \cdots + (2n - 1) = n^2$$

—Nicomachus of Gerasa (circa A. D. 100)

Sums of Odd Integers II

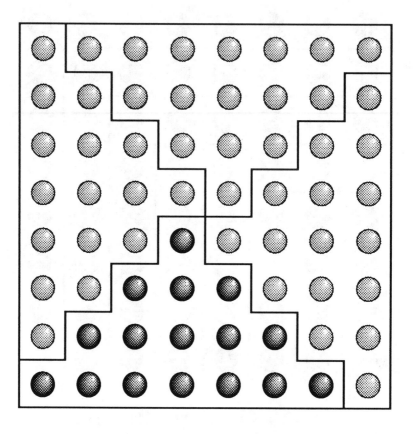

$$1 + 3 + \cdots + (2n-1) = \frac{1}{4}(2n)^2 = n^2$$

Sums of Odd Integers III

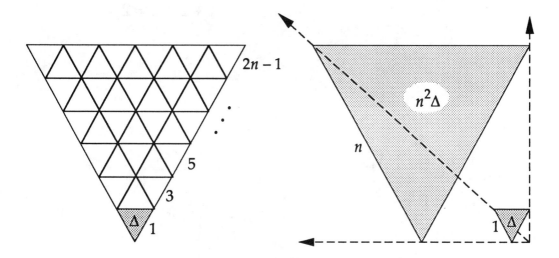

$$\Delta + 3{\cdot}\Delta + \cdots + (2n-1){\cdot}\Delta = A = n^2{\cdot}\Delta$$

$$\sum_{i=1}^{n} (2i-1) = n^2$$

—Jenő Lehel

Squares and Sums of Integers

I.

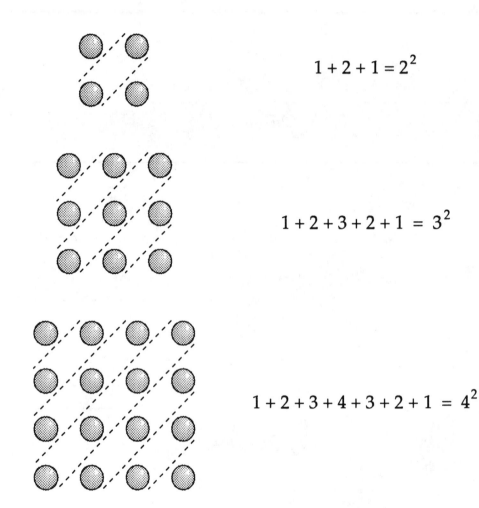

$$1 + 2 + 1 = 2^2$$

$$1 + 2 + 3 + 2 + 1 = 3^2$$

$$1 + 2 + 3 + 4 + 3 + 2 + 1 = 4^2$$

$$1 + 2 + \cdots + (n-1) + n + (n-1) + \cdots + 2 + 1 = n^2$$

—"The ancient Greeks"
(as cited by Martin Gardner)

II.

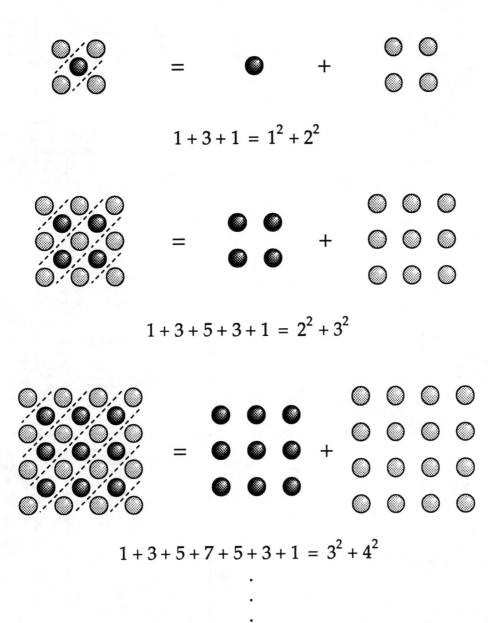

$$1 + 3 + 1 = 1^2 + 2^2$$

$$1 + 3 + 5 + 3 + 1 = 2^2 + 3^2$$

$$1 + 3 + 5 + 7 + 5 + 3 + 1 = 3^2 + 4^2$$

.
.
.

$$1 + 3 + \cdots + (2n-1) + (2n+1) + (2n-1) + \cdots + 3 + 1 = n^2 + (n+1)^2$$

—Hee Sik Kim

Arithmetic Progressions with Sum Equal to the Square of the Number of Terms

$$\sum_{k=n}^{3n-2} k = (2n-1)^2; \quad n = 1, 2, 3, \cdots.$$

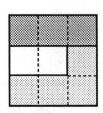

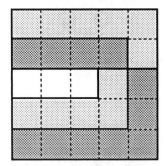

$n = 4$

$4 + 5 + 6 + 7 + 8 + 9 + 10 = 7^2$

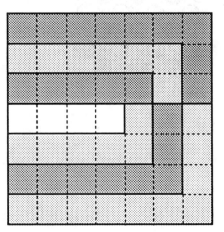

—James O. Chilaka

Sums of Squares I

$$1^2 + 2^2 + \cdots + n^2 = \frac{1}{3}\,n(n+1)(n+\tfrac{1}{2})$$

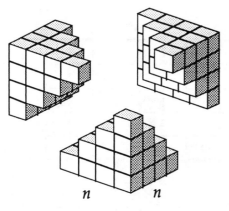

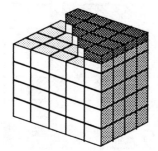

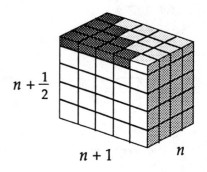

—Man-Keung Siu

Sums of Squares II

$$3(1^2 + 2^2 + \cdots + n^2) = (2n + 1)(1 + 2 + \cdots + n)$$

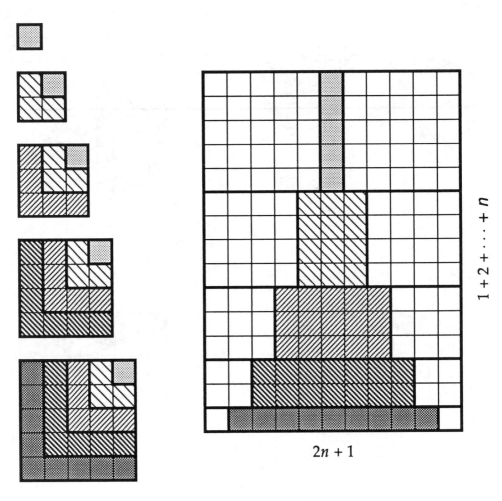

$2n + 1$

$1 + 2 + \cdots + n$

—Martin Gardner and Dan Kalman
(independently)

Sums of Squares III

$$3(1^2 + 2^2 + \cdots + n^2) = \frac{1}{2}n(n + 1)(2n + 1)$$

n	n	$\ldots$	n	n		n	$n-1$	$\ldots$	2	1		1	2	$\ldots$	$n-1$	n
$n-1$	$n-1$	$\ldots$	$n-1$			n	$n-1$	$\ldots$	2			2	3	$\ldots$	n	
$\cdot$	$\cdot$		$\cdot$			$\cdot$	$\cdot$		$\cdot$			$\cdot$	$\cdot$			
$\cdot$	$\cdot$		$\cdot$	$+$		$\cdot$	$\cdot$		$\cdot$	$+$		$\cdot$	$\cdot$		$\cdot$	
$\cdot$	$\cdot$	$\cdot$				$\cdot$	$\cdot$	$\cdot$				$\cdot$	$\cdot$			
2	2					n	$n-1$					$n-1$	n			
1						n						n				

$$=$$

$2n+1$	$2n+1$	$\ldots$	$2n+1$	$2n+1$
$2n+1$	$2n+1$	$\ldots$	$2n+1$	
$\cdot$	$\cdot$	$\cdot$		
$\cdot$	$\cdot$	$\cdot$		
$\cdot$	$\cdot$	$\cdot$		
$2n+1$	$2n+1$			
$2n+1$				

—Sidney H. Kung

Sums of Squares IV

$$\sum_{k=1}^{n} k^2 = \left(\sum_{k=1}^{n} k\right)^2 - 2\sum_{k=1}^{n-1}\left[\left(\sum_{i=1}^{k} i\right)(k+1)\right]$$

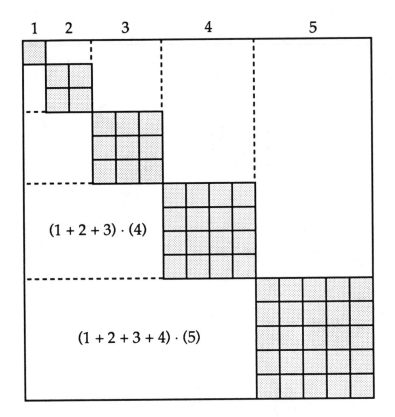

—James O. Chilaka

Sums of Squares V

$$\sum_{i=1}^{n}\sum_{j=i}^{n} j = \sum_{i=1}^{n} i^2$$

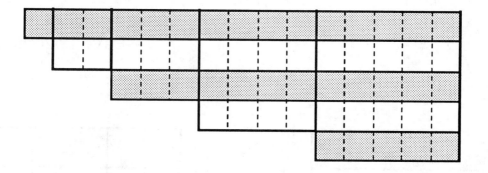

—Pi-Chun Chuang

Alternating Sums of Squares

I.

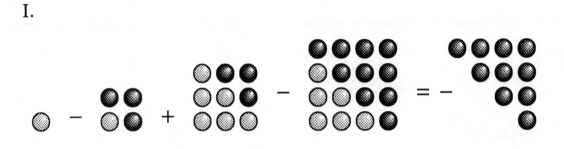

$$\sum_{k=1}^{n} (-1)^{k+1} k^2 = (-1)^{n+1} T_n = (-1)^{n+1} \frac{n(n+1)}{2}$$

—Dave Logothetti

II.

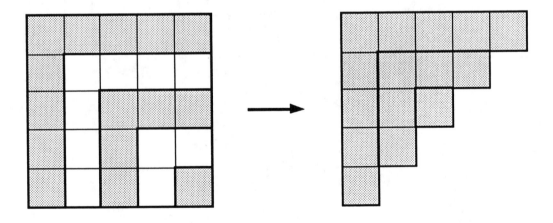

$$n^2 - (n-1)^2 + \cdots + (-1)^{n-1}(1)^2 = \sum_{k=0}^{n} (-1)^k (n-k)^2 = \frac{n(n+1)}{2}$$

—Steven L. Snover

Sums of Squares of Fibonacci Numbers

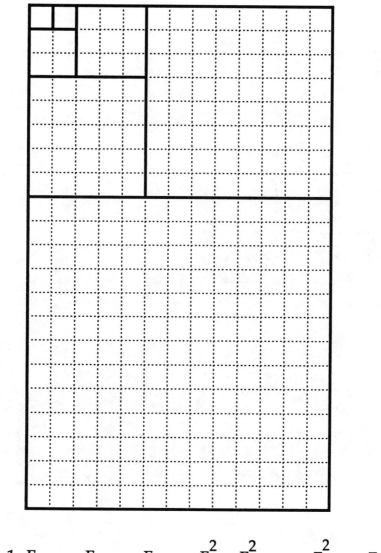

$$F_1 = F_2 = 1;\ F_{n+2} = F_{n+1} + F_n \Rightarrow F_1^2 + F_2^2 + \ldots + F_n^2 = F_n F_{n+1}$$

—Alfred Brousseau

Sums of Cubes I

$$1^3 + 2^3 + 3^3 + \cdots + n^3 = (1 + 2 + 3 + \cdots + n)^2$$

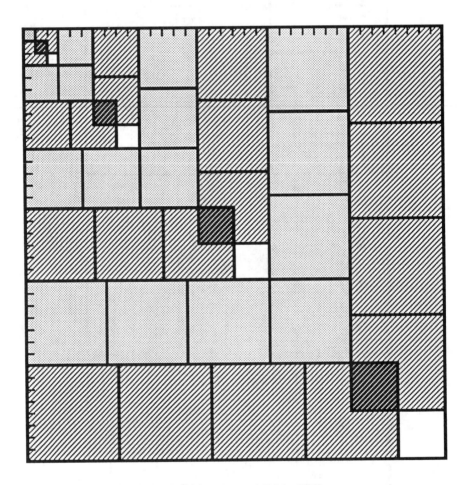

—Solomon W. Golomb

Sums of Cubes II

$$1^3 + 2^3 + 3^3 + \cdots + n^3 = (1 + 2 + 3 + \cdots + n)^2$$

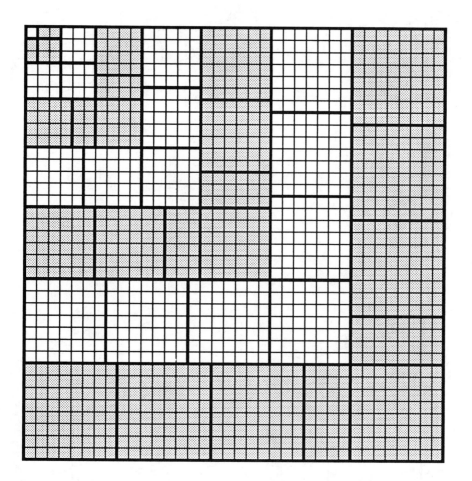

—J. Barry Love

Sums of Cubes III

$$1^3 + 2^3 + 3^3 + \cdots + n^3 = (1 + 2 + 3 + \cdots + n)^2$$

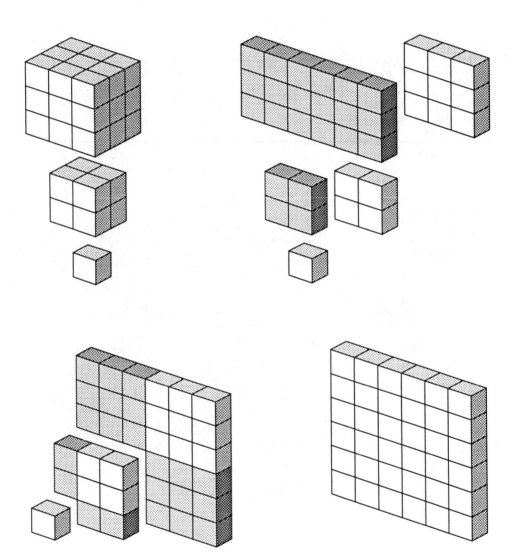

—Alan L. Fry

Sums of Cubes IV

$$1^3 + 2^3 + 3^3 + \cdots + n^3 = \frac{1}{4}[n(n+1)]^2$$

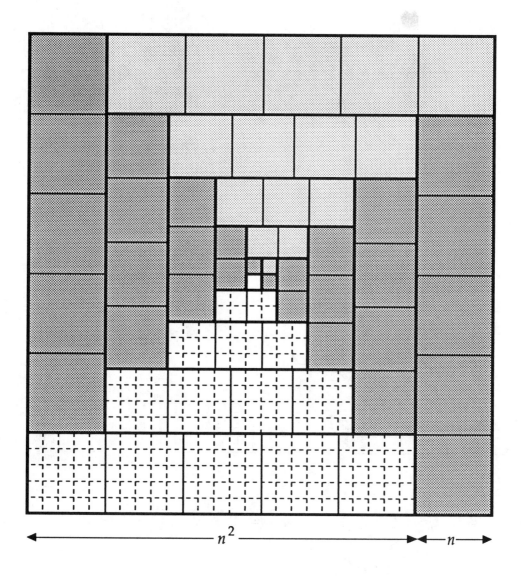

—Antonella Cupillari and Warren Lushbaugh
(independently)

Sum of Cubes V

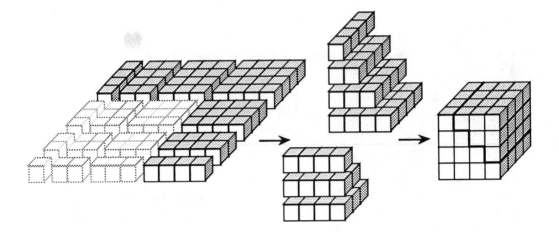

$$t_n = 1 + 2 + \cdots + n \;\;\Rightarrow\;\; t_n^2 - t_{n-1}^2 = n^3$$

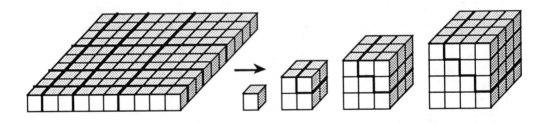

$$t_n^2 = (1 + 2 + \cdots + n)^2 = 1^3 + 2^3 + 3^3 + \cdots + n^3$$

—RBN

Sum of Cubes VI

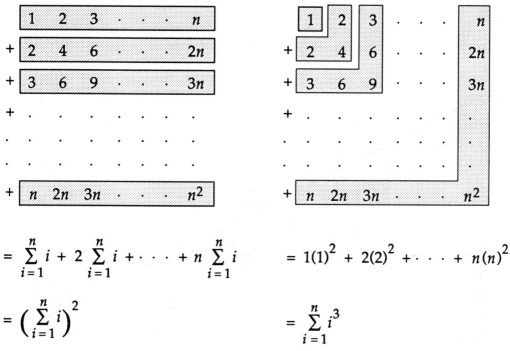

$$= \sum_{i=1}^{n} i + 2 \sum_{i=1}^{n} i + \cdots + n \sum_{i=1}^{n} i$$

$$= 1(1)^2 + 2(2)^2 + \cdots + n(n)^2$$

$$= \left(\sum_{i=1}^{n} i \right)^2$$

$$= \sum_{i=1}^{n} i^3$$

—Farhood Pouryoussefi

Sums of Integers and Sums of Cubes

$$1 + 2 + \cdots + n = \tfrac{1}{2}n(n + 1)$$

$$1^3 + 2^3 + \cdots + n^3 = \left(\tfrac{1}{2}n(n + 1)\right)^2$$

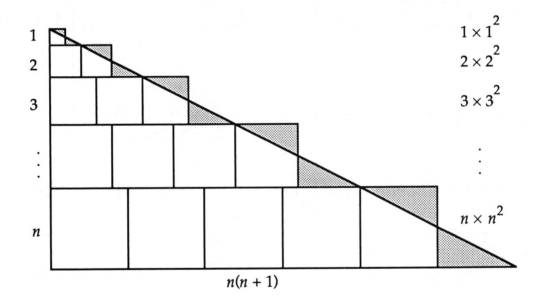

—Georg Schrage

Sums of Odd Cubes are Triangular Numbers

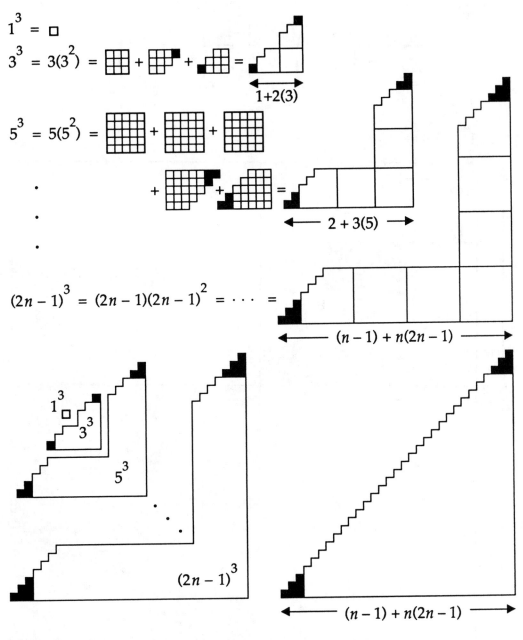

$$1^3 + 3^3 + 5^3 + \cdots + (2n-1)^3 = 1 + 2 + 3 + \cdots + (2n^2-1) = n^2(2n^2-1)$$

—Monte J. Zerger

Sums of Fourth Powers

$$\sum_{i=1}^{n} i^4 = \left(\sum_{i=1}^{n} i^2\right)^2 - 2\left[\sum_{k=2}^{n}\left(k^2 \sum_{i=1}^{k-1} i^2\right)\right]$$

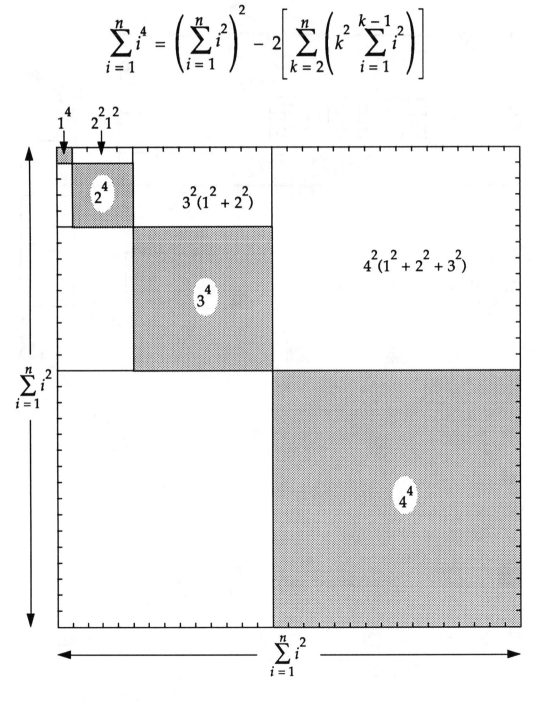

—Elizabeth M. Markham

kth Powers as Sums of Consecutive Odd Numbers

$$n^k = (n^{k-1} - n + 1) + (n^{k-1} - n + 3) + \cdots + (n^{k-1} - n + 2n - 1);$$
$$k = 2, 3, \cdots.$$

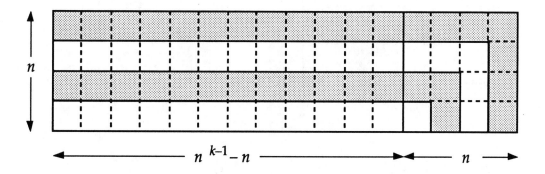

—N. Gopalakrishnan Nair

Sums of Triangular Numbers I

$$T_n = 1 + 2 + \cdots + n \Rightarrow T_1 + T_2 + \cdots + T_n = \frac{n(n+1)(n+2)}{6}$$

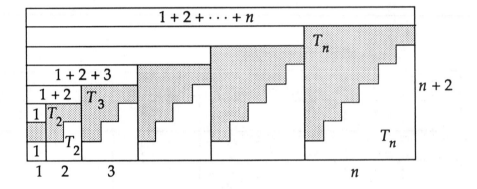

$$3(T_1 + T_2 + \cdots + T_n) = (n+2) \cdot T_n$$

$$T_1 + T_2 + \cdots + T_n = \frac{(n+2)}{3} \cdot \frac{n(n+1)}{2} = \frac{n(n+1)(n+2)}{6}$$

—Monte J. Zerger

Sums of Triangular Numbers II

$$T_k = 1 + 2 + \cdots + k \implies \sum_{k=1}^{n} T_k = \frac{1}{6} n(n+1)(n+2)$$

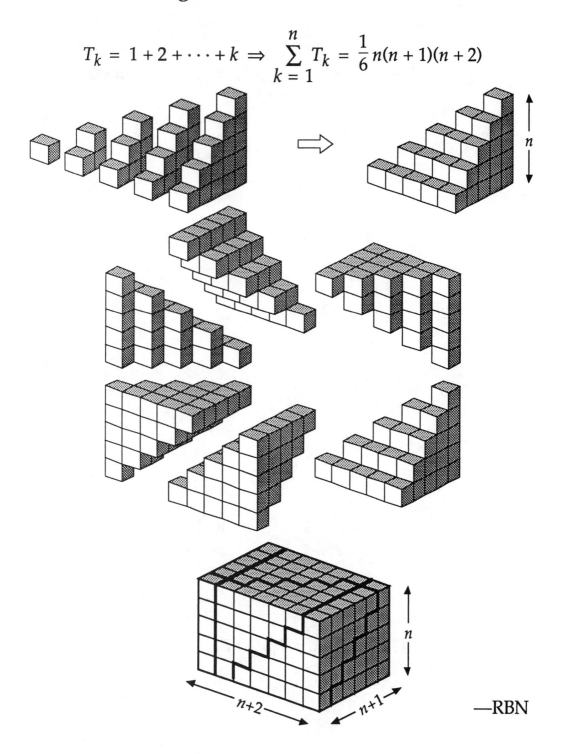

—RBN

Sums of Triangular Numbers III

$$T_k = 1 + 2 + \cdots + k \Rightarrow 3 \sum_{k=1}^{n} T_k = \frac{1}{2}n(n+1)(n+2)$$

```
1                         1                         n
1 2                       2   1                     n−1 n−1
1 2 3                     3   2   1                 n−2 n−2 n−2
· · ·      ·        +     ·    ·    ·    ·     +     ·    ·      ·   ·
· · ·      ·              ·    ·    ·    ·           ·    ·      ·       ·
1 2   ···  n−1          n−1 n−2   ···   1           2    2     ···      2
1 2   ···  n−1 n        n   n−1   ···   2 1         1    1     ···      1 1
```

```
              n+2
              n+2 n+2
              n+2 n+2 n+2
       =       ·    ·       ·
                 ·    ·    ·    ·
              n+2 n+2     ···    n+2
              n+2 n+2     ···    n+2 n+2
```

$$3(T_1 + T_2 + \ldots + T_n) = T_n \cdot (n+2)$$

Sums of Oblong Numbers I

$$(1 \times 2) + (2 \times 3) + (3 \times 4) + \cdots + (n-1)n = \frac{(n-1)n(n+1)}{3}$$

(i)

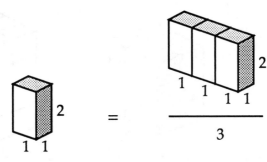

(ii)

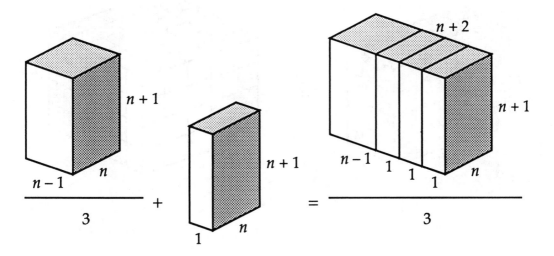

—T. C. Wu

Sums of Oblong Numbers II

$$3(1{\cdot}2 + 2{\cdot}3 + 3{\cdot}4 + \cdots + n(n + 1)) = n(n + 1)(n + 2)$$

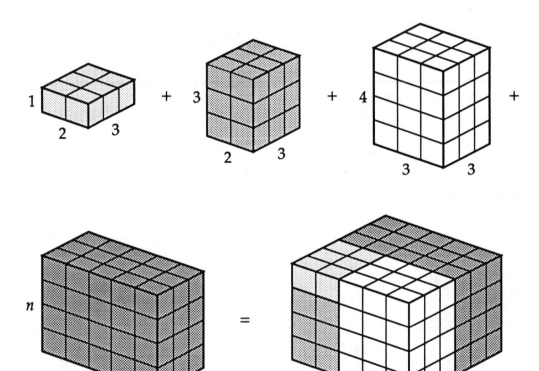

—Sidney H. Kung

Sums of Oblong Numbers III

$$(1 \times 2) + (2 \times 3) + \cdots + (n-1) \times n = \frac{1}{3}[n^3 - n]$$

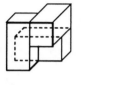 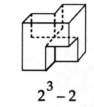

$3(1 \times 2)$ $=$ $2^3 - 2$

 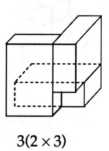 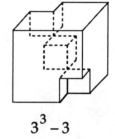

$3(1 \times 2)$ $+$ $3(2 \times 3)$ $=$ $3^3 - 3$

 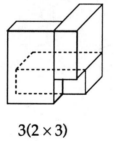 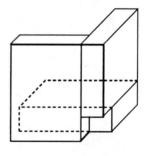

$3(1 \times 2)$ $+$ $3(2 \times 3)$ $+$ $3(3 \times 4)$ $=$

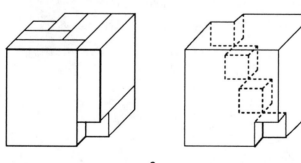

 $4^3 - 4$

—Ali R. Amir-Moéz

Sums of Pentagonal Numbers

$$\frac{1 \cdot 2}{2} + \frac{2 \cdot 5}{2} + \frac{3 \cdot 8}{2} + \cdots + \frac{n(3n-1)}{2} = \frac{n^2(n+1)}{2}$$

—William A. Miller

On Squares of Positive Integers

$$T_n = 1 + 2 + \cdots + n \quad \Rightarrow$$

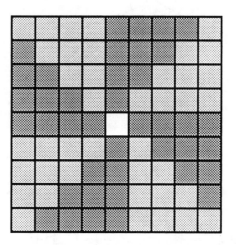

$$(2n + 1)^2 = 8T_n + 1$$

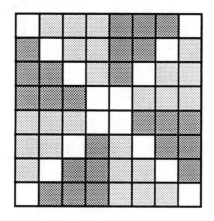

$$(2n)^2 = 8T_{n-1} + 4n$$

—Edwin G. Landauer

Consecutive Sums of Consecutive Integers

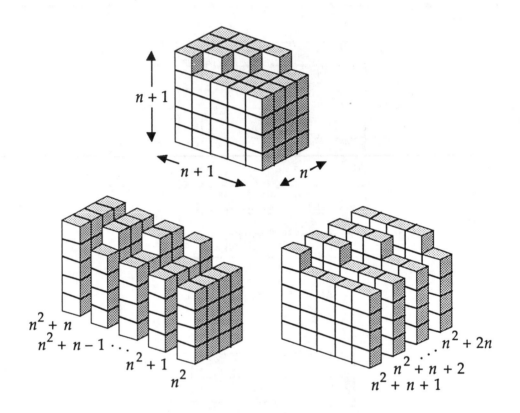

$$1 + 2 = 3$$
$$4 + 5 + 6 = 7 + 8$$
$$9 + 10 + 11 + 12 = 13 + 14 + 15$$
$$16 + 17 + 18 + 19 + 20 = 21 + 22 + 23 + 24$$

$$\vdots$$

$$n^2 + (n^2 + 1) + \cdots + (n^2 + n) = (n^2 + n + 1) + \cdots + (n^2 + 2n)$$

—RBN

Count the Dots

$$\sum_{k=1}^{n} k + \sum_{k=1}^{n-1} k = n^2$$

$$\sum_{k=1}^{n} k + n^2 = \sum_{k=n+1}^{2n} k$$

—Warren Page

Identities for Triangular Numbers

$$T_n = 1 + 2 + \cdots + n \quad \Rightarrow$$

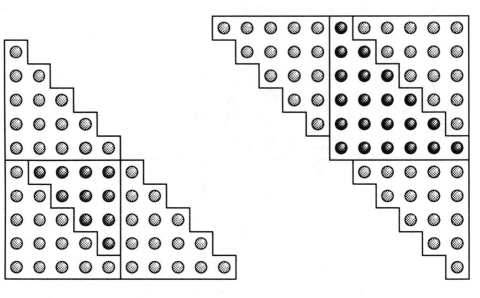

$$3T_n + T_{n-1} = T_{2n} \qquad\qquad 3T_n + T_{n+1} = T_{2n+1}$$

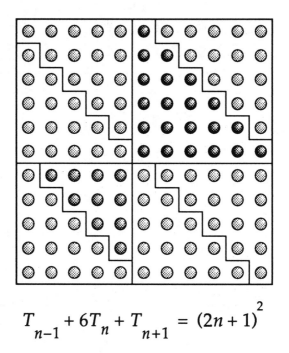

$$T_{n-1} + 6T_n + T_{n+1} = (2n+1)^2$$

A Triangular Identity

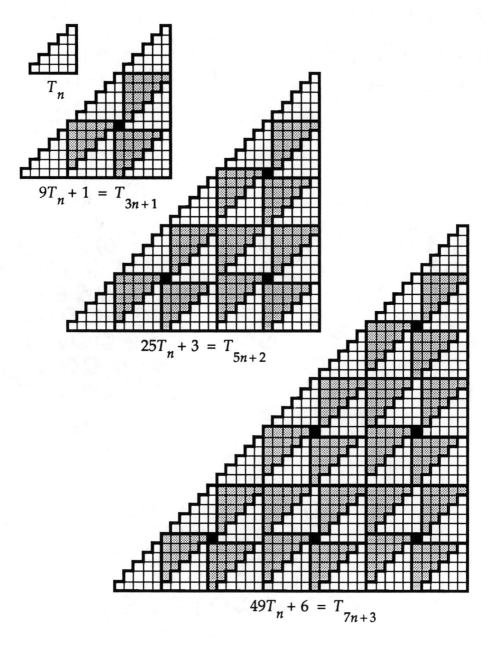

T_n

$$9T_n + 1 = T_{3n+1}$$

$$25T_n + 3 = T_{5n+2}$$

$$49T_n + 6 = T_{7n+3}$$

$$T_n = 1 + 2 + \cdots + n \Rightarrow (2k+1)^2 T_n + T_k = T_{(2k+1)n+k}$$

—RBN

Every Hexagonal Number is a Triangular Number

$$H_n = 1+5 +\cdots+(4n-3) \atop T_n = 1+2 +\cdots+n \Bigg\} \Rightarrow H_n = 3T_{n-1} + T_n = T_{2n-1} = n(2n-1)$$

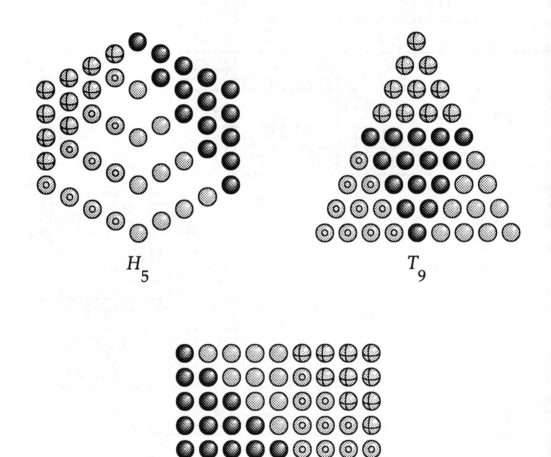

H_5 T_9

$5 \cdot 9$

One Domino = Two Squares:
Concentric Squares

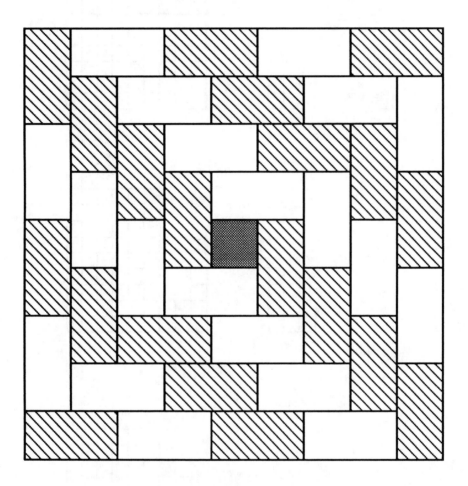

$$1 + 4 \cdot 2 + 8 \cdot 2 + 12 \cdot 2 + 16 \cdot 2 = 9^2$$

$$1 + 2 \sum_{k=1}^{n} 4k = (2n + 1)^2$$

—Shirley A. Wakin

Sums of Consecutive Powers of Nine are Sums of Consecutive Integers

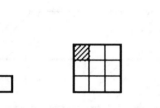

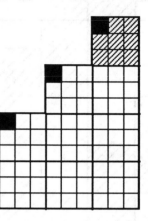

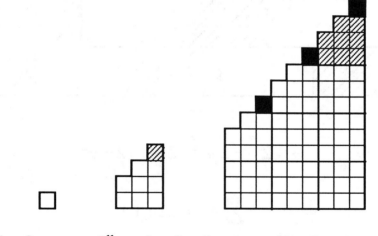

$$1 + 9 + \cdots + 9^n = 1 + 2 + 3 + \cdots + (1 + 3 + \cdots + 3^n)$$

—RBN

Sums of Hex Numbers Are Cubes

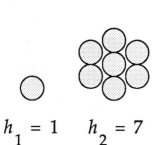
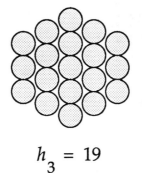

$$h_1 = 1 \qquad h_2 = 7 \qquad h_3 = 19 \qquad h_4 = 37$$

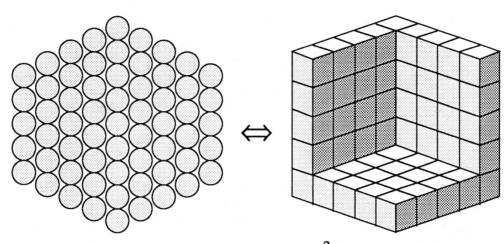

$$\Longleftrightarrow$$

$$h_n = n^3 - (n-1)^3$$

$$\therefore\ h_1 + h_2 + \cdots h_n = n^3.$$

Every Cube is the Sum of Consecutive Odd Numbers

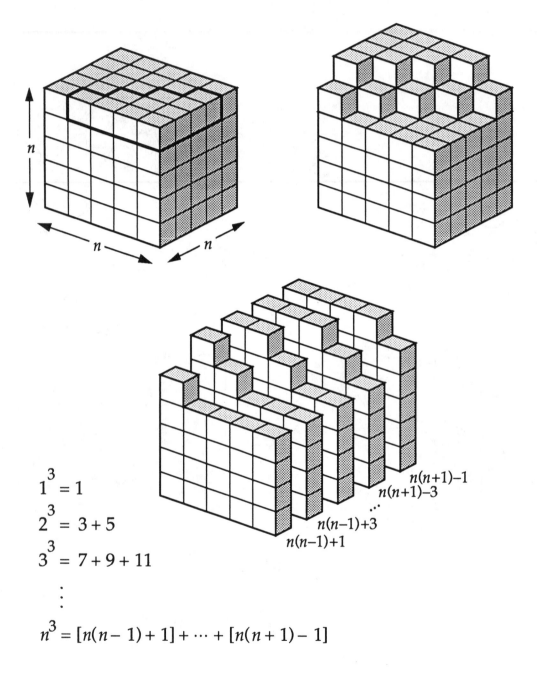

$1^3 = 1$

$2^3 = 3 + 5$

$3^3 = 7 + 9 + 11$

$\vdots$

$n^3 = [n(n-1) + 1] + \cdots + [n(n+1) - 1]$

—RBN

The Cube as an Arithmetic Sum

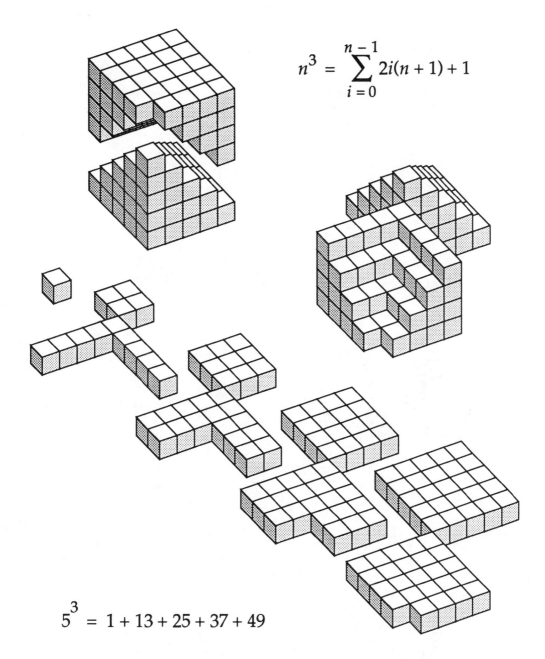

$$n^3 = \sum_{i=0}^{n-1} 2i(n+1) + 1$$

$$5^3 = 1 + 13 + 25 + 37 + 49$$

—Robert Bronson and
Christopher Brueningsen

Sequences & Series

On a Property of the Sequence of Odd Integers (Galileo, 1615)

$$\frac{1}{3} = \frac{1+3}{5+7} = \frac{1+3+5}{7+9+11} = \cdots .$$

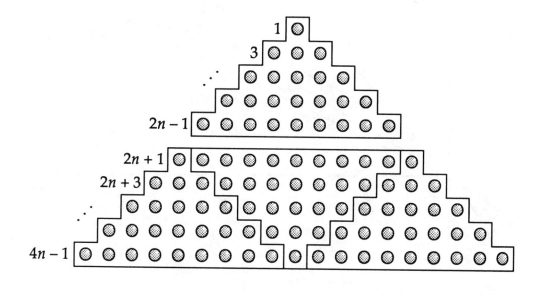

$$\frac{1+3+\ldots+(2n-1)}{(2n+1)+(2n+3)+\ldots+(4n-1)} = \frac{1}{3}$$

REFERENCE

S. Drake, *Galileo Studies*, The University of Michigan Press, Ann Arbor, 1970, pp. 218-219.

—RBN

A Monotone Sequence Bounded by e

$$\forall n \geq 1, \left(1 + \frac{1}{n}\right)^n < \left(1 + \frac{1}{n+1}\right)^{n+1} < e.$$

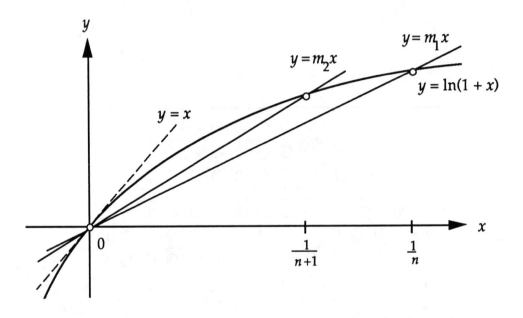

$$n \geq 1 \Rightarrow m_1 < m_2 < 1$$

$$\Rightarrow \frac{\ln(1 + \frac{1}{n})}{\frac{1}{n}} < \frac{\ln(1 + \frac{1}{n+1})}{\frac{1}{n+1}} < 1$$

$$\Rightarrow \left(1 + \frac{1}{n}\right)^n < \left(1 + \frac{1}{n+1}\right)^{n+1} < e$$

—RBN

A Recusively Defined Sequence for e

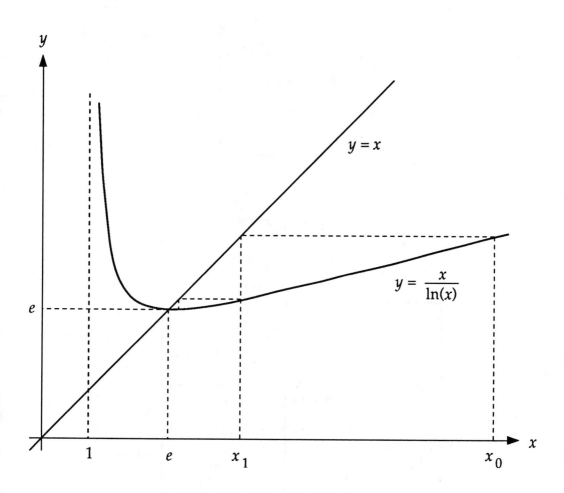

$$x_0 > 1 \ \& \ x_{n+1} = \frac{x_n}{\ln(x_n)} \Rightarrow \lim x_n = e$$

—Thomas P. Dence

Geometric Sums

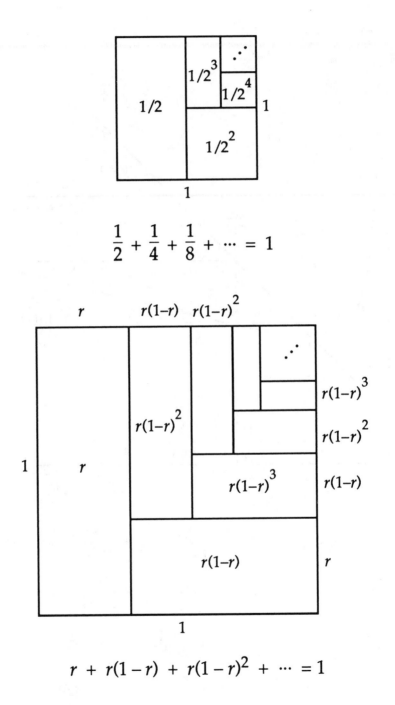

$$\frac{1}{2} + \frac{1}{4} + \frac{1}{8} + \cdots = 1$$

$$r + r(1-r) + r(1-r)^2 + \cdots = 1$$

—Warren Page

Geometric Series I

$$\sum_{n=0}^{\infty} ar^n = \frac{a}{1-r}$$

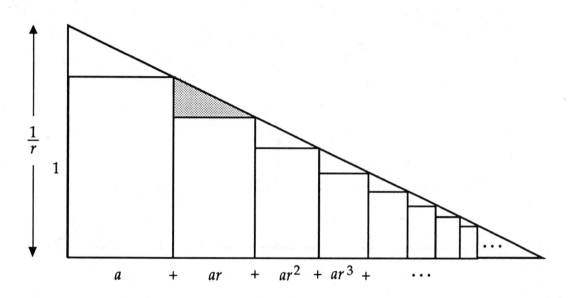

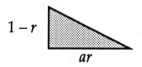

$$\frac{a + ar + ar^2 + ar^3 + \cdots}{1/r} = \frac{ar}{1-r}$$

—J. H. Webb

Geometric Series II

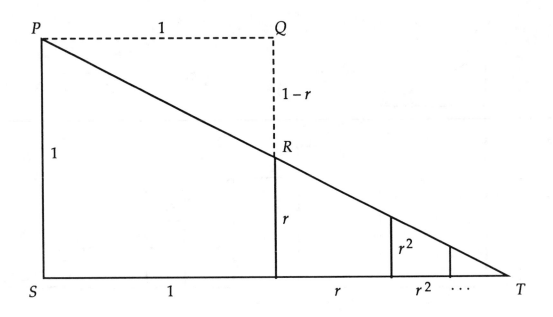

$$\Delta PQR \approx \Delta TSP$$

$$\therefore 1 + r + r^2 + \dots = \frac{1}{1 - r}.$$

—Benjamin G. Klein and Irl C. Bivens

Geometric Series III

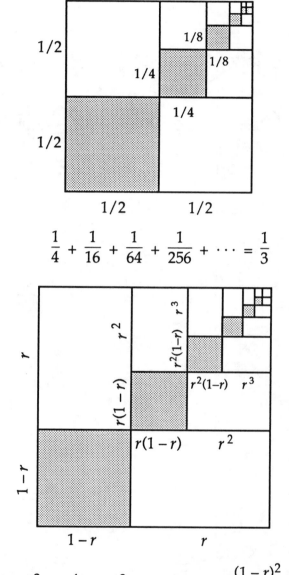

$$\frac{1}{4} + \frac{1}{16} + \frac{1}{64} + \frac{1}{256} + \cdots = \frac{1}{3}$$

$$(1-r)^2 + r^2(1-r)^2 + r^4(1-r)^2 + \ldots = \frac{(1-r)^2}{(1-r)^2 + 2r(1-r)} = \frac{1-r}{1+r}$$

$$1 + r^2 + r^4 + \ldots = \frac{1}{1-r^2}$$

$$a + ar + ar^2 + \ldots = \frac{a}{1-r}$$

—Sunday A. Ajose

Geometric Series IV

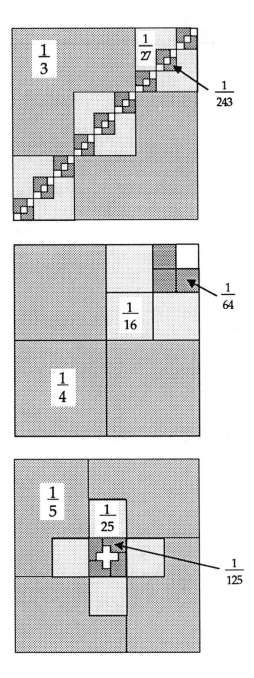

$$2\left(\frac{1}{3} + 3 \cdot \frac{1}{27} + 9 \cdot \frac{1}{243} + \cdots\right) = 1$$

$$2\sum_{n=1}^{\infty}\frac{1}{3^n} = 1$$

$$\sum_{n=1}^{\infty}\frac{1}{3^n} = \frac{1}{2}$$

$$3\sum_{n=1}^{\infty}\frac{1}{4^n} = 1$$

$$\sum_{n=1}^{\infty}\frac{1}{4^n} = \frac{1}{3}$$

$$4\sum_{n=1}^{\infty}\frac{1}{5^n} = 1$$

$$\sum_{n=1}^{\infty}\frac{1}{5^n} = \frac{1}{4}$$

—Elizabeth M. Markham

Gabriel's Staircase

$$\sum_{k=1}^{\infty} kr^k = \frac{r}{(1-r)^2} \text{ for } 0 < r < 1$$

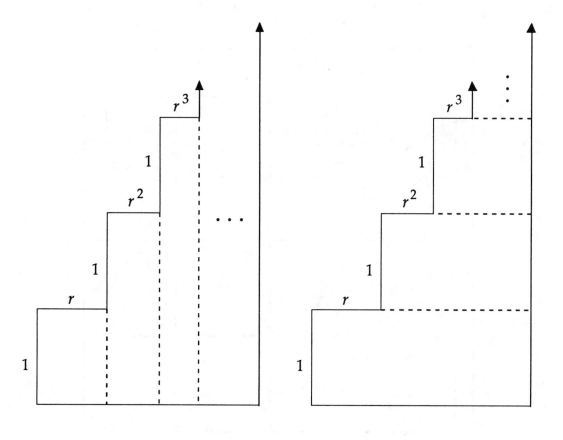

$$\sum_{k=1}^{\infty} kr^k = \sum_{k=1}^{\infty}\sum_{i=k}^{\infty} r^i = \frac{r}{(1-r)^2}$$

—Stuart G. Swain

Differentiated Geometric Series

$$1 + 2\left(\frac{1}{2}\right) + 3\left(\frac{1}{4}\right) + 4\left(\frac{1}{8}\right) + \cdots = 4$$

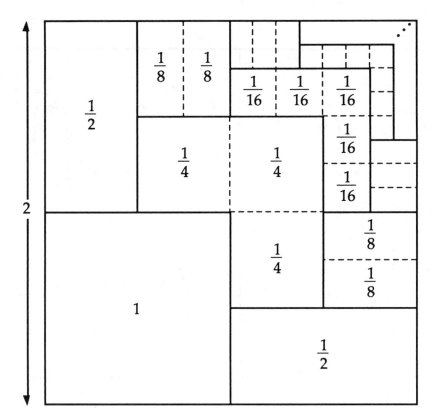

$$1 + 2r + 3r^2 + 4r^3 + \ldots = \left(\frac{1}{1-r}\right)^2, \; 0 \le r < 1$$

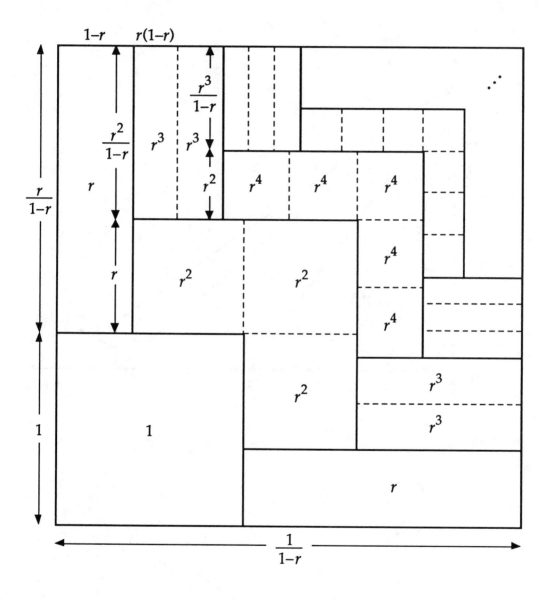

—RBN

$$\frac{1}{1 \cdot 2} + \frac{1}{2 \cdot 3} + \cdots + \frac{1}{n(n+1)} = \frac{n}{n+1}$$

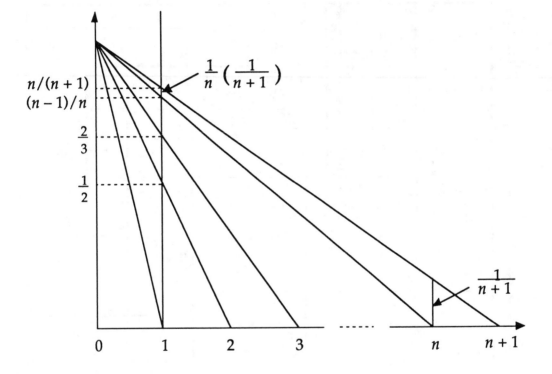

—Roman W. Wong

The Series of Reciprocals of Triangular Numbers

$$\frac{1}{1} + \frac{1}{3} + \frac{1}{6} + \ldots + \frac{1}{\binom{n+1}{2}} + \ldots = 2$$

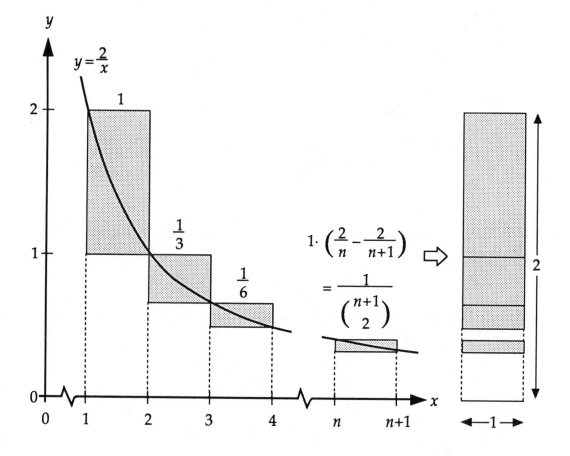

—RBN

The Alternating Harmonic Series

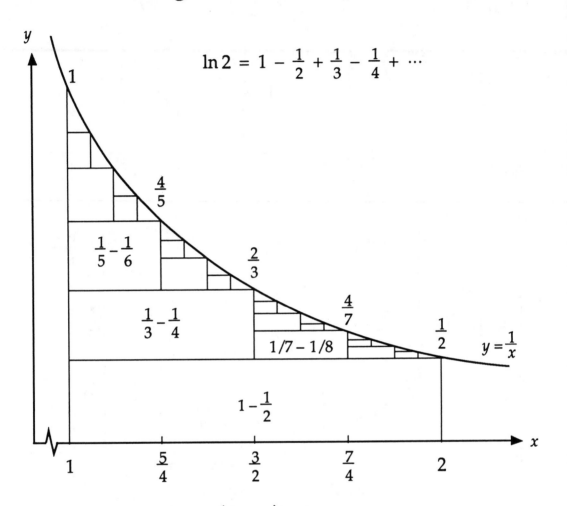

$$\ln 2 = 1 - \frac{1}{2} + \frac{1}{3} - \frac{1}{4} + \cdots$$

$$y = \frac{1}{x}$$

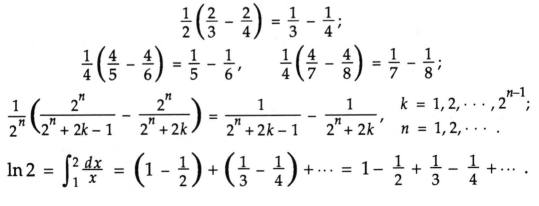

$$\frac{1}{2}\left(\frac{2}{3} - \frac{2}{4}\right) = \frac{1}{3} - \frac{1}{4};$$

$$\frac{1}{4}\left(\frac{4}{5} - \frac{4}{6}\right) = \frac{1}{5} - \frac{1}{6}, \qquad \frac{1}{4}\left(\frac{4}{7} - \frac{4}{8}\right) = \frac{1}{7} - \frac{1}{8};$$

$$\frac{1}{2^n}\left(\frac{2^n}{2^n + 2k - 1} - \frac{2^n}{2^n + 2k}\right) = \frac{1}{2^n + 2k - 1} - \frac{1}{2^n + 2k}, \quad \begin{array}{l} k = 1, 2, \cdots, 2^{n-1}; \\ n = 1, 2, \cdots. \end{array}$$

$$\ln 2 = \int_1^2 \frac{dx}{x} = \left(1 - \frac{1}{2}\right) + \left(\frac{1}{3} - \frac{1}{4}\right) + \cdots = 1 - \frac{1}{2} + \frac{1}{3} - \frac{1}{4} + \cdots.$$

—Mark Finkelstein

$$\sin(2n+1)\theta = \sin\theta + 2\sin\theta \sum_{k=1}^{n} \cos 2k\theta$$

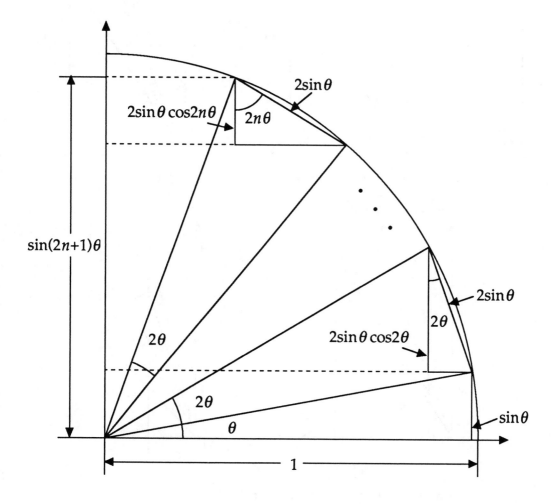

—J. Chris Fisher and E. L. Koh

An Arctangent Identity and Series

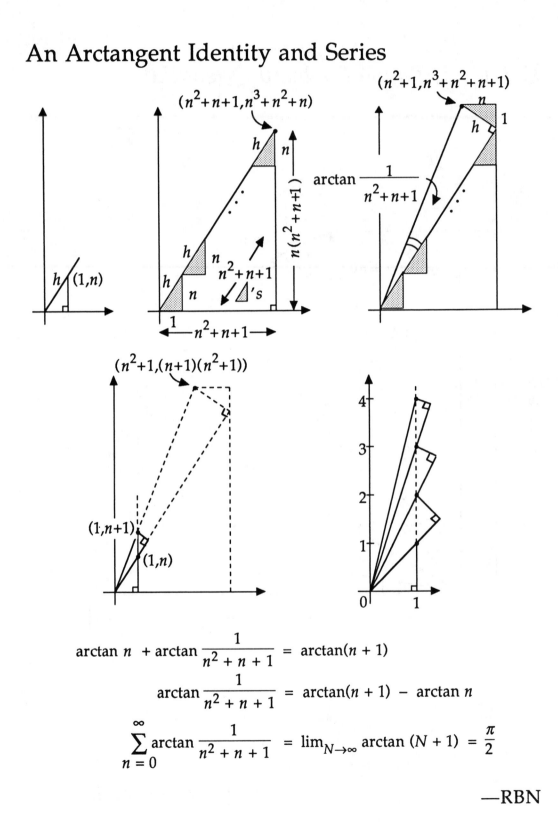

$$\arctan n + \arctan \frac{1}{n^2 + n + 1} = \arctan(n + 1)$$

$$\arctan \frac{1}{n^2 + n + 1} = \arctan(n + 1) - \arctan n$$

$$\sum_{n=0}^{\infty} \arctan \frac{1}{n^2 + n + 1} = \lim_{N \to \infty} \arctan(N + 1) = \frac{\pi}{2}$$

—RBN

Miscellaneous

A 2 × 2 Determinant is the Area of a Parallelogram

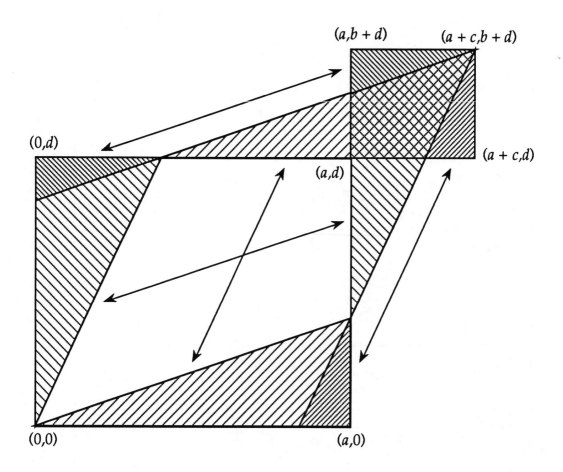

$$\begin{vmatrix} a & b \\ c & d \end{vmatrix} = ad - bc = \left\| \Box \right\| - \left\| \Box \right\| = \left\| \diagup \right\|$$

—Solomon W. Golomb

Area of the Parallelogram Determined by

Vectors (a,b) and $(c,d) = \pm \begin{vmatrix} a & b \\ c & d \end{vmatrix} = \pm(ad - bc)$

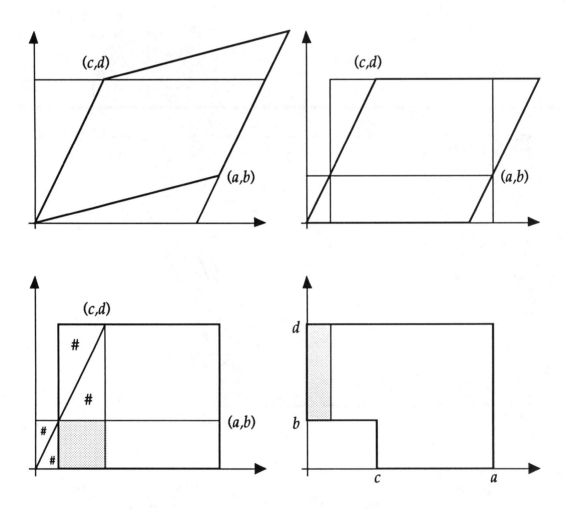

—Yihnan David Gau

The Characteristic Polynomials of AB and BA are Equal

$$-\lambda^n |AB - \lambda I| = \left| \begin{pmatrix} A & AB - \lambda I \\ \lambda I & 0 \end{pmatrix} \right| = \left| \begin{pmatrix} A & I \\ \lambda I & B \end{pmatrix} \begin{pmatrix} I & B \\ 0 & -\lambda I \end{pmatrix} \right| = \left| \begin{matrix} A & I \\ \lambda I & B \end{matrix} \right| (-\lambda)^n$$

$$-\lambda^n |BA - \lambda I| = \left| \begin{pmatrix} 0 & \lambda I \\ BA - \lambda I & \lambda B \end{pmatrix} \right| = \left| \begin{pmatrix} A & I \\ \lambda I & B \end{pmatrix} \begin{pmatrix} -I & 0 \\ A & \lambda I \end{pmatrix} \right| = \left| \begin{matrix} A & I \\ \lambda I & B \end{matrix} \right| (-\lambda)^n$$

—Sidney H. Kung

The Gaussian Quadrature as the Area of Either Trapezoid

$$\tfrac{1}{2}(b - a)(f(\overline{a}) + f(\overline{b})) = \tfrac{1}{2}(b - a)(h(a) + h(b))$$

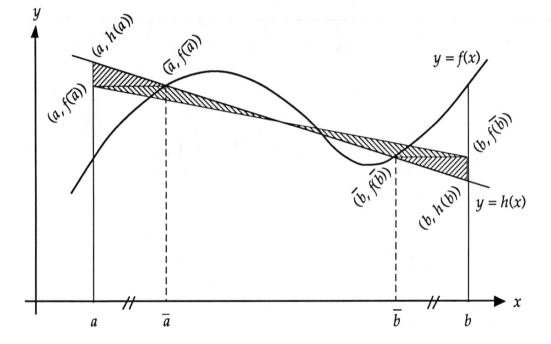

—Mike Akerman

Inductive Construction of an Infinite Chessboard with Maximal Placement of Nonattacking Queens

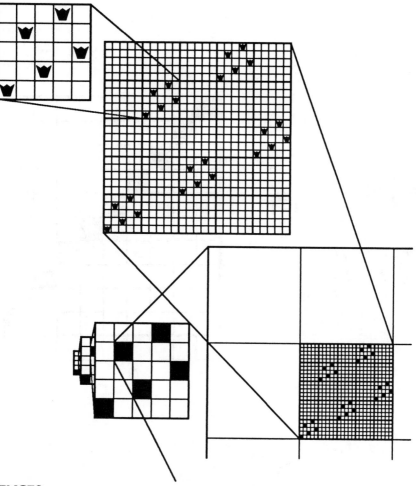

REFERENCES

1. Dean S. Clark and Oved Shisha, Invulnerable Queens on an Infinite Chessboard, *Annals of the New York Academy of Sciences, The Third International Conference on Combinatorial Mathematics,* 1989, 133-139.
2. M. Kraitchik, *La Mathématique des Jeux ou Récréations Mathématiques,* Imprimerie Stevens Frères, Bruxelles, 1930, 349-353.

—Dean S. Clark and Oved Shisha

Combinatorial Identities

$$\binom{n}{2} = \frac{1}{2}(n^2 - n) = \sum_{i=1}^{n-1} i$$

$$\binom{n+1}{2} = \binom{n}{2} + n$$

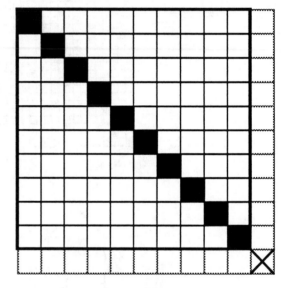

—James O. Chilaka

$$3 \sum_{j=0}^{n} \binom{3n}{3j} = 8^n + 2(-1)^n, \text{ by Inclusion-Exclusion in}$$

Pascal's Triangle

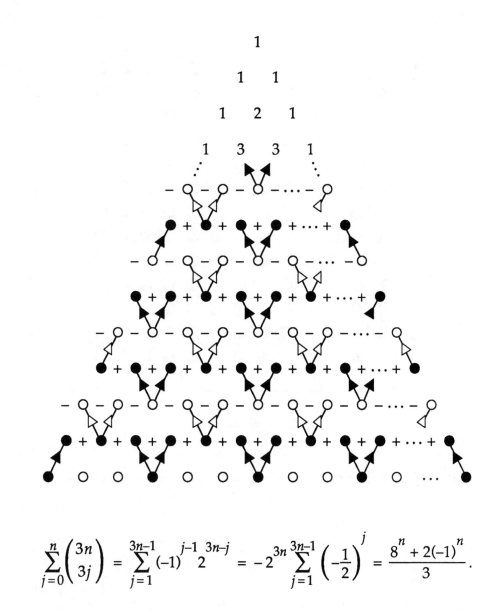

$$\sum_{j=0}^{n} \binom{3n}{3j} = \sum_{j=1}^{3n-1} (-1)^{j-1} 2^{3n-j} = -2^{3n} \sum_{j=1}^{3n-1} \left(-\frac{1}{2}\right)^j = \frac{8^n + 2(-1)^n}{3}.$$

—Dean S. Clark

The Existence of Infinitely Many Primitive Pythagorean Triples

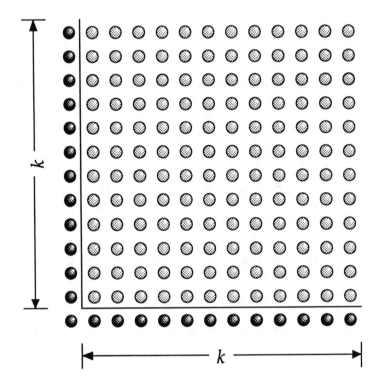

$$n^2 = 2k + 1 \implies k^2 + n^2 = (k+1)^2 \ \& \ (k, k+1) = 1$$

—Charles Vanden Eynden

Pythagorean Triples via Double Angle Formulas

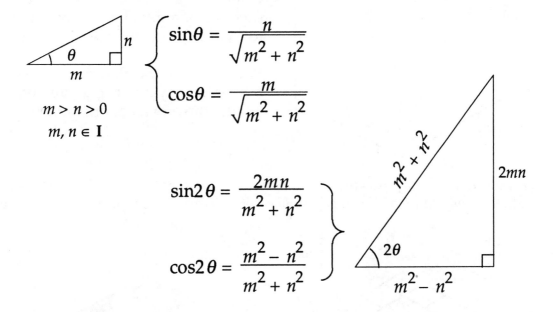

$$\sin\theta = \frac{n}{\sqrt{m^2 + n^2}}$$

$$\cos\theta = \frac{m}{\sqrt{m^2 + n^2}}$$

$$m > n > 0$$

$$m, n \in \mathbf{I}$$

$$\sin 2\theta = \frac{2mn}{m^2 + n^2}$$

$$\cos 2\theta = \frac{m^2 - n^2}{m^2 + n^2}$$

—David Houston

The Problem of the Calissons

A *calisson* is a French sweet that looks like two equilateral triangles meeting along an edge. Calissons could come in a box shaped like a regular hexagon, and their packing would suggest an interesting combinatorial problem. Suppose a box with side of length n is filled with sweets of sides of length 1. The short diagonal of each calisson in the box is parallel to a pair of sides of the box.

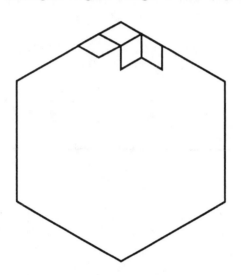

We refer to these three possibilities by saying that a calisson admits three distinct orientations.

THEOREM: *In any packing, the number of calissons with a given orientation is one-third of the total number of calissons in the box.*

PROOF:

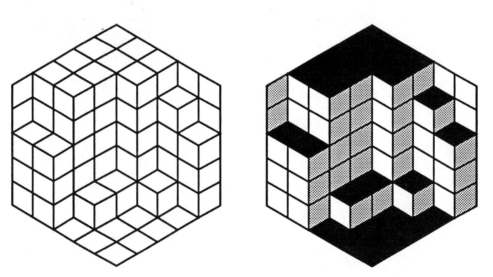

—Guy David and Carlos Tomei

Recursion

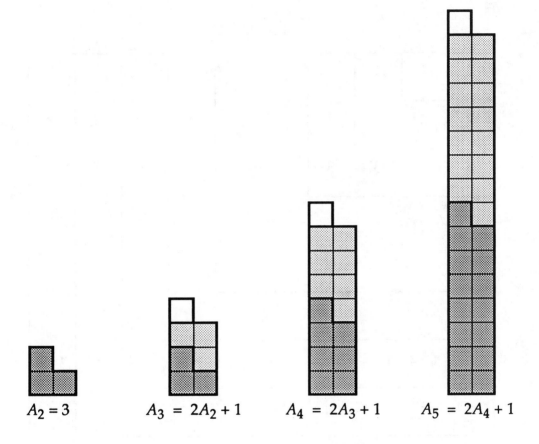

$A_2 = 3$ $A_3 = 2A_2 + 1$ $A_4 = 2A_3 + 1$ $A_5 = 2A_4 + 1$

$$A_2 = 3 \ \& \ A_n = 2A_{n-1} + 1 \ \Leftrightarrow \ A_n = 2(2^{n-1}) - 1 = 2^n - 1$$

—Shirley Wakin

$$\prod_{k=1}^{n} k^{k} \cdot k! = (n!)^{n+1}$$

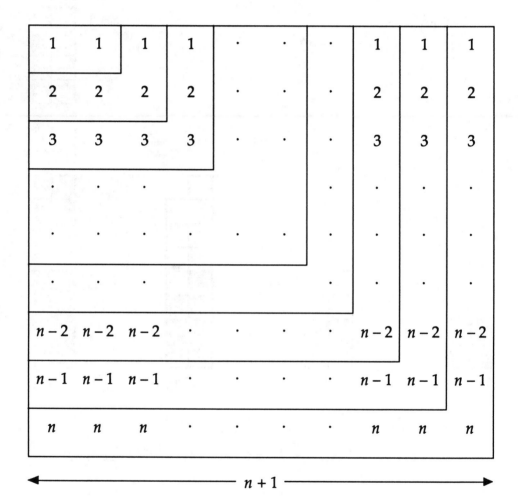

—Edward T. H. Wang

Sources

Trigonometry, Calculus & Analytic Geometry

Inequalities

page	source

Inequalities (continued)

61   II. *Mathematics Magazine*, in press.
62   *Mathematics Magazine*, in press.
63   *Mathematics Magazine*, vol. 66, no. 1 (Feb. 1993), p. 65.
64   *Mathematics Magazine*, in press.
65   *Mathematics Magazine*, in press.
66   *College Mathematics Journal*, vol. 24, no. 2 (March 1993). p. 165.

Integer Sums

69   *Scientific American*, vol. 229, no. 4 (Oct. 1973), p. 114.
70   *Mathematics Magazine*, vol. 57, no. 2 (March 1984), p. 104.
71   Reprinted by permission from *Historical Topics for the
      Mathematics Classroom*, p. 54, author Bernard H. Gundlach,
      copyright © 1969 by the National Council of Teachers of
      Mathematics, Inc.
73   *Mathematics Magazine*, vol. 64, no. 2 (April 1991), p. 103.
74   *Scientific American*, vol 229, no. 4 (Oct. 1973), p. 115.
75   *Mathematics Magazine*, vol. 66, no. 3 (June 1993), p. 166.
76   *Mathematics Magazine*, vol. 59, no. 2 (April 1986), p. 92.
77   *Mathematics Magazine*, vol. 57, no. 2 (March 1984), p. 92.
78   *Scientific American*, vol. 229, no. 4 (Oct. 1973), p. 115.
      College Mathematics Journal, vol. 22, no. 2 (March 1991), p. 124.
79   *College Mathematics Journal*, vol. 20, no. 3 (May 1989), p. 205.
80   *Mathematics Magazine*, vol. 56, no. 2 (March 1983), p. 90.
81   *College Mathematics Journal*, vol. 20, no. 2 (March 1989), p. 123.
82   I.   *Mathematics Magazine*, vol. 60, no. 5 (Dec. 1987), p. 291.
      II. *Mathematics Magazine*, vol. 65, no. 2 (April 1992), p. 90.
83   M. Bicknell & V. E. Hoggatt, Jr. (eds.), *A Primer for the Fibonacci
      Numbers*, The Fibonacci Association, San Jose, 1972, p. 147.
84   *Mathematical Gazette*, vol. 49, no. 368 (May 1965), p. 199.
85   *Mathematics Magazine*, vol. 50, no. 2 (March 1977), p. 74.
86   *Mathematics Magazine*, vol. 58, no. 1 (Jan. 1985), p. 11.
87   *Mathematics Magazine*, vol. 62, no. 4 (Oct. 1989), p. 259.
      Mathematical Gazette, vol. 49, no. 368 (May 1965), p. 200.
88   *Mathematics Magazine*, vol. 63, no. 3 (June 1990), p. 178.
89   *Mathematics Magazine*, vol. 62, no. 5 (Dec. 1989), p. 323.
90   *Mathematics Magazine*, vol. 65, no. 3 (June 1992), p. 185.

Index of Names

Technical Note

The manuscript for this book was edited and printed using Microsoft® Word 5.1 on an Apple® Macintosh™ IIfx computer. The graphics were produced in Claris™ MacDraw® Pro 1.5v1. The text is set in the Palatino font, with special characters in the Symbol font. Many of the displayed equations were produced with MacΣqn™ 3.0 from Software for Recognition Technologies. The manuscript was printed on an Apple® Laserwriter® II NTX.